Andrea Schleicher

Überlebensstrategien in dynamischen Lebensräumen

Andrea Schleicher

Überlebensstrategien in dynamischen Lebensräumen

Die Anwendbarkeit der Metagemeinschaftstheorie auf funktionelle Pflanzengruppen

Südwestdeutscher Verlag für Hochschulschriften

Impressum/Imprint (nur für Deutschland/only for Germany)
Bibliografische Information der Deutschen Nationalbibliothek: Die Deutsche Nationalbibliothek verzeichnet diese Publikation in der Deutschen Nationalbibliografie; detaillierte bibliografische Daten sind im Internet über http://dnb.d-nb.de abrufbar.
Alle in diesem Buch genannten Marken und Produktnamen unterliegen warenzeichen-, marken- oder patentrechtlichem Schutz bzw. sind Warenzeichen oder eingetragene Warenzeichen der jeweiligen Inhaber. Die Wiedergabe von Marken, Produktnamen, Gebrauchsnamen, Handelsnamen, Warenbezeichnungen u.s.w. in diesem Werk berechtigt auch ohne besondere Kennzeichnung nicht zu der Annahme, dass solche Namen im Sinne der Warenzeichen- und Markenschutzgesetzgebung als frei zu betrachten wären und daher von jedermann benutzt werden dürften.

Verlag: Südwestdeutscher Verlag für Hochschulschriften GmbH & Co. KG
Dudweiler Landstr. 99, 66123 Saarbrücken, Deutschland
Telefon +49 681 37 20 271-1, Telefax +49 681 37 20 271-0
Email: info@svh-verlag.de

Zugl.: Oldenburg: Carl-v.Ossietzky-Universität Oldenburg, Diss., 2010

Herstellung in Deutschland:
Schaltungsdienst Lange o.H.G., Berlin
Books on Demand GmbH, Norderstedt
Reha GmbH, Saarbrücken
Amazon Distribution GmbH, Leipzig
ISBN: 978-3-8381-2621-0

Imprint (only for USA, GB)
Bibliographic information published by the Deutsche Nationalbibliothek: The Deutsche Nationalbibliothek lists this publication in the Deutsche Nationalbibliografie; detailed bibliographic data are available in the Internet at http://dnb.d-nb.de.
Any brand names and product names mentioned in this book are subject to trademark, brand or patent protection and are trademarks or registered trademarks of their respective holders. The use of brand names, product names, common names, trade names, product descriptions etc. even without a particular marking in this works is in no way to be construed to mean that such names may be regarded as unrestricted in respect of trademark and brand protection legislation and could thus be used by anyone.

Publisher: Südwestdeutscher Verlag für Hochschulschriften GmbH & Co. KG
Dudweiler Landstr. 99, 66123 Saarbrücken, Germany
Phone +49 681 37 20 271-1, Fax +49 681 37 20 271-0
Email: info@svh-verlag.de

Printed in the U.S.A.
Printed in the U.K. by (see last page)
ISBN: 978-3-8381-2621-0

Copyright © 2011 by the author and Südwestdeutscher Verlag für Hochschulschriften GmbH & Co. KG and licensors
All rights reserved. Saarbrücken 2011

Contents

1 Preface 3
2 Metacommunity theory and the functional trait concept: Towards a more general understanding of biodiversity 11
    2.1 Metacommunity theory 11
    2.2 The functional trait concept 20
    2.3 Box 1: Assembly Rules 28
    2.4 Reconciling metacommunity ecology and the functional trait concept 31
3 Study design 37
    3.1 Study area 37
    3.2 Sampling design: Bridging scales 40
    3.3 Functional diversity 43
4 Seed number and terminal velocity determine plant response to habitat connectivity in an urban landscape 51
    4.1 Introduction 52
    4.2 Methods 55
    4.3 Results 60
    4.4 Discussion 63
5 Effects of the resident community on colonizing plants: A functional approach 75
    5.1 Introduction 77
    5.2 Methods 79
    5.3 Results 87
    5.4 Discussion 92
6 Functional patterns during succession: Is plant community assembly trait-driven? 101
    6.1 Introduction 103
    6.2 Methods 107
    6.3 Results 114
    6.4 Discussion 124
7 Synthesis 133
    7.1 Relationships between functional traits and assembly filters 134
    7.2 Plant functional groups in the study area 146
    7.3 Relevance of metacommunity paradigms in the study area 148

	7.4	Predicting the relevance of metacommunity paradigms in plant communities 154
	7.5	A glance at methods 157
	7.6	Future research needs 161
	7.7	Outlook 163
8	Appendix 167
9	References 183

Chapter 1

"... community ecology is a mess, with so much contingency that useful generalisations are hard to find"

John H. Lawton

1 Preface

Over the last century the world has seen a surge of extinctions without parallel in pre-human periods (see Pimm *et al.* 1995; Woodruff 2001; Wake & Vredenburg 2008). Even if accurate predictions are difficult their magnitude alone is alarming. Expecting 3,000–30,000 losses per year, 50% of all animal and plant species worldwide could have disappeared by the end of the present century (Wilson 2002).

Concern on future species losses has animated efforts to understand the mechanisms that generate and maintain biodiversity patterns over space and time. How are communities assembled? Although this question has been at the heart of community ecology for almost a century (e.g. Clements 1916; Tansley 1917), the general principles or „rules" behind community assembly remain to be elucidated. The outcome of community assembly is highly contingent on the organisms involved and their environment, which has hampered the development of rules applicable to more than a single or a small number of species (Lawton 1999; McGill *et al.* 2006). The lack of more general rules, that allow the prediction of the community composition of specific sites in the landscape, has been called the „agony" of community ecology (Lewontin 1974; Weiher & Keddy 1999). Yet, while some ecologists have questioned if community ecology will ever be able to produce general principles (Simberloff 2004), major insights arose from two ecological disciplines: metacommunity ecology and the functional trait concept.

Community ecology has traditionally focused on local mechanisms, and adopted only recently the view that processes at coarse scales, too, govern biodiversity. Dispersal between local populations can, for instance, explain the fact, that the number of species coexisting in a set of communities connected by dispersal exceeds the number of species in any single community (Wilson 1992). The study of dispersal processes has a long-standing tradition in animal ecology, and triggered the development of metacommunity theory: a theory that explicitly deals with the effects of dispersal for biodiversity at fine and coarse scales. However, the relevance of metacommunity dynamics for plants is still subject to debate (e.g. Freckleton & Watkinson 2002; Ehrlén & Eriksson 2003; Freckleton & Watkinson 2003), although sowing experiments have proven the commonness of dispersal limitation in

CHAPTER 1

plant communities: Species may be missing in a local community simply because they have not yet managed to arrive (Tilman 1997; Turnbull *et al.* 2000).

The need to integrate metacommunity ideas into plant ecological studies becomes the more urgent as habitat loss and fragmentation are considered main drivers of biodiversity loss (Vitousek 1994; Pimm & Raven 2000). Understanding the ways in that dispersal among communities influences plant diversity is thus central for predicting the consequences of changes in the spatiotemporal configuration of habitats following fragmentation.

A second promising avenue is the functional trait concept. Based on the observation that plants with similar ecological properties often share distinct physio-morphological features or traits, community ecologists have hypothesized that, vice versa, the response of a given species to an external force may be predictable from its functional traits. Accordingly, predictions are not only possible for single species but for groups of species with similar traits, so-called functional groups. Due to its focus on functional differences rather than taxonomic identities, the functional trait approach may even be used to predict the response of species across different geographic regions. As an example, recurrent patterns of leaf traits associated with the investments of nutrients and leaf dry mass have been demonstrated along global climatic gradients (Wright *et al.* 2004b).

The functional trait concept may also be used to answer the question of deterministic versus neutral community assembly. To explain biodiversity patterns deterministic concepts typically refer to differences among species regarding theirs environmental requirements and competitive rankings (Tansley 1917; Diamond 1975): Any given species distribution and abundance in a given place in space and time is predictable from its ability to thrive under the local abiotic conditions and the competitive effects exerted by neighbors. Thereby, competitive effects means that fitness components of individuals are negatively affected due to resource depletion by neighbors (Violle *et al.* 2009). Competition may be responsible if a plant does not occur under conditions that actually fit its environmental requirements or "fundamental niche". Therefore, the „realized niche" of a species is often only a small subset of its „fundamental niche" (Hutchinson 1957; Byer 1969).

Contrary to deterministic concepts, neutral concepts question the relevance of niches. More generally, they assume net ecological equality among species and even individuals, turning community composition the result of largely stochastic processes (Bell 2001; Hubbell 2001). Although neutral concepts have a long standing history in ecology (e.g. Skellam 1951; Caswell 1976), the debate about deterministic versus neutral assembly was reanimated when Stephen Hubbell published his „Unified Neutral Theory" (Hubbell 2001). In this work, Hubbell assumes that species are identical in their per capita probabilities of giving birth, dying, migrating and speciating, such as local communities are „....*structured entirely by ecological drift, random migration and random speciation*" (Hubbell 2001, p. 6). Hubbell's neutral theory was very successful in predicting local biodiversity and species abundances in a range of communities (reviewed in Hubbell 2001), thus questioning the usefulness of deterministic concepts.

Functional group approaches might be superior to species-based approaches in answering the question of deterministic versus neutral community assembly. Due to their focus on functional differences rather than taxonomic units, they are able to reveal recurring patterns in situations when species-based approaches fail (e.g. Messier *et al.* 2010). Consequently, functional group approaches may not only help to yield more general principles underlying community assembly, but also contribute to answer the question in how far community assembly is predictable from species properties at all (Lavorel & Garnier 2002; McGill *et al.* 2006).

CHAPTER 1
Outline

Although metacommunity theory and the functional trait concept have received considerable attention over the last decade, rarely have they been discussed in an integrated way. This might, however, decidedly promote the identification of general rules underlying community assembly, which has been a major aim of community ecology ever since. In the present work, ideas arising from metacommunity theory and the functional trait concept were put together to investigate the mechanisms that drive communities of vascular plants. The main goal was to answer two questions:

- Is plant community assembly trait-driven or trait-neutral, i.e. is community composition predictable from functional traits or do neutral forces prevail?
- How do dispersal processes influence biodiversity, and what may be concluded on the relevance of metacommunity dynamics for the spatial and temporal dynamics of plant communities?

Thereto, main aspects of metacommunity theory and the functional trait concept are presented in Chapter 2. Metacommunity theory and its principal paradigms are introduced, and general obstacles to its applicability to natural plant communities are extracted. Another part of Chapter 2 is dedicated to the functional group approach. From a plant perspective, the theoretical background and major current issues of this concept are summarized. Subsequently, metacommunity theory and the functional trait concept are synthesized to derive research questions. In Chapter 3, study area and design are presented. It is shown how the study design answered central methodological challenges related to the empirical evaluation of main drivers of community assembly.

In the subsequent three chapters, the methods and findings of three specific questions are described. Chapter 4 addresses the role of dispersal limitation. It is demonstrated in how far plant species distributions depended on the spatiotemporal configuration of habitats, and if there was a relationship between the degree to that plants were dispersal limited and their functional traits. In Chapter 5, the role of dispersal limitation was omitted. Seed bank composition was analyzed to determine the importance of local environmental conditions and competition for structuring plant communities at the colonization stage. Chapter 6 explores the functional composition of communities to draw conclusions on the relevance of trait-driven versus trait neutral assembly.

PREFACE

In Chapter 7, these results are synthesized and discussed in respect to the applicability of metacommunity paradigms to the community dynamics of the study area. Methodological difficulties are summarized and fields of future research extracted. Finally, implications of the assembly mechanisms identified in the study area are interpreted in respect to biodiversity conservation and biodiversity response to environmental changes.

This work is a presentation of my original research work including the collection of data in the field, their processing, analysis, and the writing of manuscript drafts. Exceptions are the Chapters 4, 5 and 6, which were prepared together with co-authors acknowledged in the respective chapters.

Chapter 2

"When an ecologist says "there goes a badger" he should include in his thoughts some definite idea of the animal's place in the community to which it belongs, just as if he had said "there goes the vicar." "

Charles Elton

2 Metacommunity theory and the functional trait concept: Towards a more general understanding of biodiversity

For considerable time, traditional community ecology and spatial ecology have been conducted in almost mutually exclusive ways because of a focus at different scales. While traditional community ecology attempted to understand biodiversity patterns from local abiotic and biotic factors ("niche-based view": Ozinga 2008), spatial ecology concentrated on dispersal processes at a landscape scale. In this "dispersal-based view", populations or communities in small „islands" of suitable habitat depend on colonization from a large „mainland source". Species may be missing in some islands simply because they have not (yet) been able to colonize, which makes dispersal an important structuring force of biodiversity (MacArthur & Wilson 1967). Later, metapopulation theory enlarged this idea by exploring the effects of dispersal among local habitats without the assumption of a mainland source (Levins 1969). Biodiversity patterns are then the consequence of permanent colonization and extinction events. But dispersal-based concepts neglect local processes. For instance, successful colonization requires that the locality's environmental conditions fit the colonizer's niche requirements. If habitat suitability is contingent on the presence of other mobile species, complex dynamics may be the consequence. Such dynamics are conceptualized in metacommunity theory.

2.1 Metacommunity theory

A metacommunity may be defined as a set of local communities of interacting species which are linked by dispersal (Hanski & Gilpin 1991; Wilson 1992), and a community as a collection of species occupying a particular locality or habitat (Holyoak et al. 2005). Metacommunity theory distinguishes several ways, how dispersal processes can drive community composition over space and time. They have been categorized into four main paradigms (see Figure 1) (Leibold et al. 2004; Holyoak et al. 2005).

CHAPTER 2

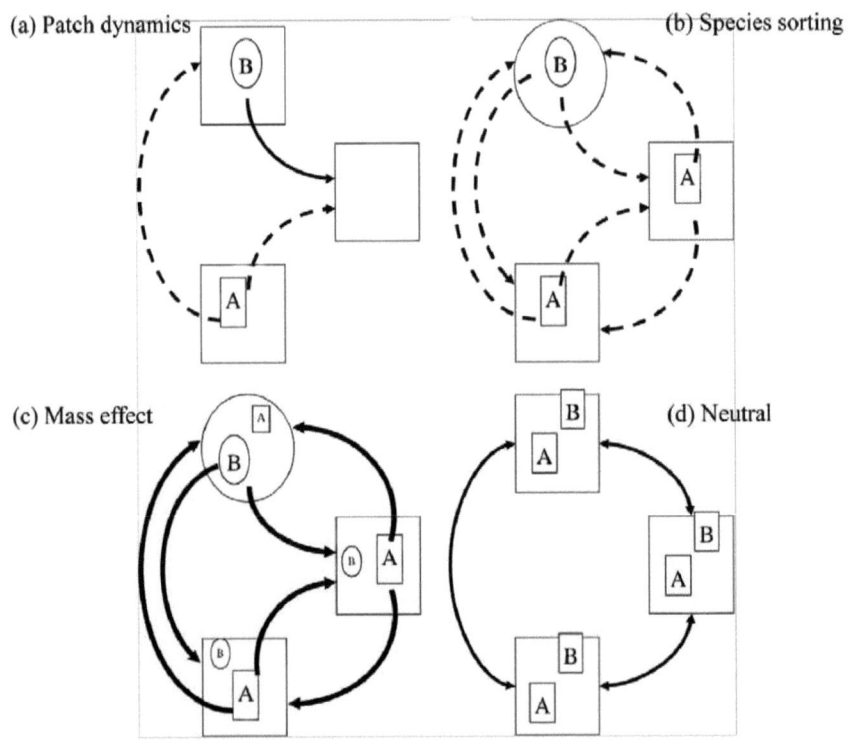

Figure 1: Schematic representation of the four paradigms of metacommunity theory for two competing species with populations A and B. Arrows connect source populations with potential colonization sites, shown as large boxes or ovals. Solid arrows indicate higher dispersal than dashed arrows and either unidirectional movement (single-headed arrows) or bidirectional movement (double-headed arrows). The degree to which a species is the competitive superior in a site is shown by the matching of the smaller box or oval (denoting its habitat requirements) with the site symbol. The four paradigms illustrated are (a) patch dynamics, (b) species sorting, (c) mass effects and (d) neutral. In (a) the patch dynamics paradigm is shown with conditions that permit coexistence: a competition-colonization trade-off is illustrated with species A being a superior competitor but species B being a superior colonist; the third patch is vacant and could become occupied by either species. In (b) species are separated into spatial niches and dispersal is not sufficient to alter their distribution. In (c) mass effects cause species to be present in both source and sink habitats; the smaller letters and symbols indicate smaller sized populations. In (d) all species are currently present in all patches; species would gradually be lost from the region and would be replaced by speciation. By Leibold et al. (2004).

Species sorting paradigm

The species sorting paradigm builds on traditional niche theory in that species separate along environmental gradients (Dobzhansky 1951; Elton 1972; Whittaker 1975). Environmental heterogeneity ensures that each species is the best competitor in some location along the gradient with associated effects on population fitness and the outcome of competitive interactions (Leibold 1998; Leibold *et al.* 2004).

Species sorting is conditioned on upper and lower thresholds to the levels of dispersal (Chase *et al.* 2005). A minimum level of dispersal is required to ensure that each species can colonize any locality in the long run (Law & Morton 1993). Continuous sorting takes place and, finally, results in the dominance of the one species that is best adapted to local conditions. Yet, dispersal rates must be low enough to inhibit mass effects (see below): species cannot be found outside their niches (Cottenie *et al.* 2003; Holyoak *et al.* 2005).

Mass effects paradigm

The mass effects paradigm is similar to the species sorting paradigm in that each species is competitively superior in some place along the environmental gradient. It assumes, however, that dispersal directly affects local population fitness. Differences between local densities as well as high levels of dispersal enable source–sink dynamics (Pulliam 1988): Populations with negative growth rates (sinks) directly benefit from the supply of immigrants from neighboring populations with positive growth rates (sources). In an extreme case, immigration may even rescue sink populations from extinction (rescue effect, Brown & Kodric-Brown 1977; Shmida & Wilson 1985). The temporal equivalent of rescue effects are storage effects, i.e. the formation of seed or bud banks enables a species to persist in the community during periods of habitat unsuitability.

Source-sink dynamics imply however not only that immigration can counterbalance local extinctions. Inversely can emigration rates reduce local population growth rates. According to Holyoak *et al.* (2005), mass effects describe an equalization of competitive differences at the regional scale. Although mass effects are contingent on higher dispersal rates than the species sorting paradigm, an upper limit is preconditioned to avoid that overall dispersal removes the spatial variance in fitness (Amarasekare *et al.* 2004; Holyoak *et al.* 2005).

Patch dynamics paradigm

In the patch dynamics paradigm, all localities are assumed to be equal, i.e. competitive hierarchies do not switch over space or time. In a static landscape, such a setting would doom inferior competitors to extinction because the presence of superior competitors in a locality would make that habitat unsuitable. However, if habitat dynamics create localities that are temporarily free from competition, and if the inferior competitor is a superior colonizer, it may avoid competitive exclusion at the regional scale. Habitat dynamics in combination with a trade-off between dispersal and competitive ability (colonization-competition trade-off) may thus be regarded as a spatiotemporal niche for inferior competitors (Tilman 1994). Coexistence of inferior and superior competitors is possible at the regional scale, whereas at the local scale competitive exclusion is expected.

Under the patch dynamics paradigm, local community composition should vary over time as it depends largely on spatial and temporal habitat configuration (Levin 1974; Freckleton & Watkinson 2002). Local abundances should change in a way as predicted from competitive hierarchies. Yet, the identity of the colonizing species after turnover is unpredictable, although good colonizers should at average be overrepresented (Chase *et al.* 2005).

Neutral paradigm

The neutral paradigm is the metacommunity counterpart of neutral theory. It assumes ecological equality of trophically similar species, i.e. local communities are "... *structured entirely by ecological drift, random migration and random speciation*" (Hubbell 2001, p. 6). Consequently, no predictions can be made from species identities or traits on the outcome of competition (Adler *et al.* 2007). This discerns the neutral paradigm from all other paradigms, but is particularly important when differentiating the neutral from the patch-dynamics paradigm. In both paradigms, species do not sort along environmental gradients.

In the neutral paradigm, there is no mechanism to enable long-term coexistence of a distinct set of species (Hubbell 2001). The number of coexisting species is only limited by the habitat's carrying capacity (zero-sum ecological drift). Each species is on a random walk to extinction, and dispersal is important because it counteracts or postpones local extinction. Different to all other paradigms, neutral theory further includes local speciation as an internal process to compensate for regional species losses.

The neutral paradigm cannot predict concrete compositions but only long term averages (Chase *et al.* 2005). Its usefulness is therefore often mainly seen in providing a powerful null hypothesis to test the other paradigms against (Bell 2005; Gewin 2006). In the temporal framework of ecological studies, however, its predictive value has been demonstrated for a range of taxa and ecosystems including vascular plants (reviewed in Hubbell 2001).

2.1.1 Ecological implications of metacommunity paradigms

The levels of habitat heterogeneity and dispersal form two major axes along which the four paradigms separate (Figure 2, Amarasekare *et al.* 2004; Cottenie & De Meester 2005). Habitat heterogeneity is a caveat to the species sorting and the mass effects paradigm, whereas patch dynamics and neutral paradigm assume constant or no competitive hierarchies across habitats. Among all paradigms, the mass effects paradigm requires the highest levels of dispersal and this characteristic is crucial to differ it from the species sorting paradigm. To separate patch dynamics and neutral paradigm a third axis is needed, which is provided by the relevance of ecological differences among species.

Knowledge of the prevailing paradigm is not only of theoretical interest. It has important implications for predicting biodiversity response to changes in the spatial or temporal configuration of habitats (also see Gonzalez 2005). Not only do the paradigms differ in their predictions of species bearing the greatest extinction risk. They also foresee different consequences for local and regional diversity, i.e. for single communities and for the metacommunity as a whole.

CHAPTER 2

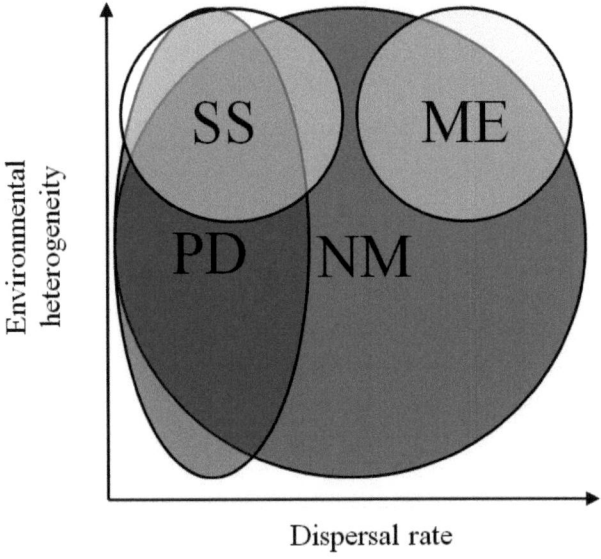

Figure 2: Separation of the four metacommunity paradigms (SS: species sorting; ME: mass effects; PD: patch dynamics; NM: neutral) along gradients of dispersal and environmental heterogeneity.

If reduced dispersal levels hinder rescue effects increased degrees of habitat isolation may involve a shift from the mass effects to the species sorting paradigm. This may first of all provoke lower species richness at the community level because smaller colonization rates are not compensated by enhanced emigration rates (Gonzalez 2005). In the long run, regional biodiversity losses may follow whereby those species should be most endangered whose populations are most dependent on the supply of immigrants from source populations.

In the species sorting paradigm, increased local species richness is expected under reduced dispersal because more species cannot reach the habitat in that they possess competitive superiority (Gonzalez 2005). This may be complemented by reduced landscape biodiversity as the number of habitat species declines with the number of niches existing in the metacommunity (Gonzalez 2005).

In the patch dynamics paradigm, reduced dispersal may slow down the colonization of superior competitors, thus indirectly favoring inferior competitors (Tilman 1997). Consequently, increased local species numbers may be expected, whereas landscape species richness should remain unaffected (Gonzalez 2005). If however the degree of dispersal limitation exceeds a threshold provided by the dispersal ability of the worst disperser (but best competitor), strong competitors may be the first to go extinct, and reduced landscape biodiversity may be predicted (Tilman *et al.* 1997; Mouquet & Loreau 2003).

For the neutral paradigm a dispersal threshold has been suggested common for all species. Biodiversity collapses as soon as the degree of dispersal limitation exceeds that threshold. Contrary to the prediction of the patch dynamics paradigm, the best colonizers might then be the first to go extinct (Solé *et al.* 2004).

2.1.2 Metacommunity theory – applicable to plants?

Knowledge of the prevailing metacommunity paradigm may be crucial to understand biodiversity patterns in space and time. But while theoretical predictions are available, empirical evidence for metacommunity dynamics in natural communities is intriguingly little (Amarasekare 2003) - and in particular for plants. Empirical studies dealing with metacommunity dynamics of plants are extremely rare. Most metacommunity studies considered communities composed of arthropods, zooplankton or microbial organisms (e.g. Jenkins 2006; Long *et al.* 2007; Louette & De Meester 2007).

As a consequence, little is known about the paradigms relevant for communities composed of vascular plants. Some ecologists generally questioned the applicability of metapopulation and metacommunity ideas to plants (see Freckleton & Watkinson 2002; Ehrlén & Eriksson 2003), although there are some well-known phenomena that point to metacommunity dynamics of plants. For instance, the occurrence of weedy species in the edges of forest fragments, resulting from the „bombardment" of fragment communities (sinks) with dispersules from the surrounding agricultural landscape (source) (Janzen 1986), may be interpreted in favor of the mass effects paradigm. The idea of competitively weak species that cannot coexist with competitively superior species within a community, but persist at the landscape scale by rapidly colonizing cleared sites (so-called "fugitive" or "transient" species: Horn & MacArthur 1972; Levin 1974), shows many similarities with the patch dynamics paradigm.

However, these phenomena have rarely been discussed in the context of metacommunity theory and its paradigms. But this is important as regional coexistence of fugitive and competitively superior species is not only in line with the patch dynamics paradigm. Elements of the neutral and the species sorting paradigm are supported if competitive abilities do not trade off with dispersal abilities, or if abiotic differences between cleared and vegetated sites are important. Indeed, stochastic processes and abiotic differences have been mentioned with reference to fugitive species before (e.g. Vitousek & Walker 1987; Lavorel *et al.* 1998; Meyer *et al.* 2009). All processes relevant in metacommunity paradigms, i.e. dispersal, species separation along environmental gradient, and neutral mechanisms, have been demonstrated to provide strong structuring forces of plant communities (e.g. Tuomisto *et al.* 2003; Levine & HilleRisLambers 2009). An integrative approach that considers the interplay of all mechanisms is needed to evaluate prevailing metacommunity dynamics, including neutral ones, for natural plant communities.

A comprehensive understanding of all mechanisms regulating plant communities as a whole is, however, hampered by the high number of species involved. Knowledge of the degree of dispersal limitation and niche differentiation is required for all species – which is an almost impossible task. Therefore, some ecologists used a multivariate variance partitioning approach (e.g. Borcard *et al.* 1992; Girdler & Barrie 2008). Variation among communities is partitioned into fractions explained by environmental descriptors, spatial descriptors, and unexplained variance. However, this approach cannot differentiate all paradigms. For instance, the importance of spatial predictors is interpreted in favor of dispersal limitation, and unexplained variance as evidence for neutral mechanisms (Chase *et al.* 2005). But dispersal limitation may be neutral or non-neutral (Clark 2008). A more reliable discrimination of patch dynamics and neutral paradigm requires further information on differences among species in their abilities to disperse and to compete – and on a trade-off between these abilities.

Conclusion

Dispersal processes possess a great potential to promote our understanding of the mechanisms that generate and maintain biodiversity. The interactive effects of dispersal and local processes are conceptualized by metacommunity theory which distinguishes four paradigms: the species sorting, mass effects, patch dynamics and the neutral paradigm.

METACOMMUNITY THEORY AND THE FUNCTIONAL TRAIT CONCEPT

However, there is little empirical validation of the relevance of alternative metacommunity paradigms for plants, and in particular in comparison to the neutral paradigm. To discuss the applicability of all four main paradigms in an integrative way, the relative importance of dispersal, environmental niche differentiation, and ecological differences among all interacting species needs to be determined. However, this seems to be hardly feasible for highly diverse, natural plant communities.

CHAPTER 2

2.2 The functional trait concept

The large number of species comprised in natural plant communities hinders the validation of metacommunity theory. It is associated with complexity which poses a major obstacle to the development of general principles (Duckworth *et al.* 2000). To move away from the description of individual cases to more general principles the use of functional groups has been advocated (Keddy 1992; Gitay & Noble 1997; McGill *et al.* 2006).

The idea of defining clusters of species with a common response to a given external force can be said to have inspired ecological research ever since. Around 300 B.C., the Greek botanist Theophrastus classified plants into trees, shrubs, and herbs (Weiher *et al.* 1999). More recent examples include the Raunkiaer's life forms, that differ in their susceptibility to frost (Raunkiaer 1934), or the C-S-R-concept, which distinguishes general plant strategies in the face of competition, environmental stress and disturbance (Grime 2002). The functional trait concept assumes that any plant's response to an external force may be predicted from its functional traits rather than requiring species-specific studies.

A trait may here be defined as any morphological, physiological or phenological feature measurable at the individual level (Violle *et al.* 2007). Yet, it becomes only functional, if the trait is directly or indirectly associated with plant performance via its effects on growth, reproduction or survival. Plant functional groups are then nonphylogenetic groupings of species that show close similarities in their response to environmental controls (Duckworth *et al.* 2000).

Regarding the many forces that possibly interfere with plant community assembly, a large number of traits should be required for a general and meaningful functional grouping. But often, several separate traits can be reduced to one or two axes of variation which capture a large proportion of the original trait variation (McGill *et al.* 2006). In a study of more than 2,500 plant species, a single axis accounted for almost 75% of the variation in six leaf traits (Wright *et al.* 2004b). This is because plant traits show a number of correlations or trade-offs, be it for evolutionary, allometric or other reasons. For instance, a plant can either generate many small seeds or few large ones (competition-fecundity trade-off, Smith & Fretwell 1974). It can allocate resources to either rapid growth or the ability to withstand physical stress (stress tolerance – growth trade-off,Loehle 1998).

Due to its focus on strategies rather than species, the functional group approach may vastly simplify plant community description (Duckworth *et al.* 2000). It can help to overcome the problem exposed by the redundancy of functionally similar species. Indeed, functional groups can show more marked responses to environmental factors than taxonomic units (for references see Weiher *et al.* 1999; Duckworth *et al.* 2000; Messier *et al.* 2010). The functional group approach also enables the formulation of principles more general than taxonomic approaches in that it can be applied to other species and ecosystems. For instance, the trait based statement „Compact plants with canopy area <30 cm² and small or absent leaves are restricted to marshes with <18 µg/g soil phosphorous" is more useful than the statement: „*Campanula aparinoides* is found only in infertile habitats"(Weiher & Keddy 1999).

2.2.1 Functional filtering view of community assembly

The main challenge in functional approaches consists in the identification of those traits that allow predicting the response of plants to the force of interest. A general framework to summarize the forces driving biodiversity patterns is the filter-view of community assembly (Keddy 1992; Lavorel & Garnier 2002). In that view, plant community assembly is understood as a hierarchy of assembly filters: Dispersal, stress and competition filter successively constrain which species from a regionally available pool of species can be found at a given locality (Figure 3). Each filter prevents the establishment of species lacking a specific combination of traits. Dispersal filtering sorts species according to their capacities to disperse, whereas traits related to stress tolerance (i.e. the ability to thrive under generally or temporally low levels of abiotic resources) and traits related to competition tolerance (i.e. the ability to exploit resources in the presence of competitors) are important in the presence of the stress and the competition filter. The species finally constituting a community are those which successfully passed all filters (Keddy 1992).

Adopting this view of community assembly, the challenge in defining ecologically meaningful functional groups boils down to identify for each filter one or more traits that capture the response of plants to that filter. In other words, traits have to be determined that govern the abilities of plants to disperse, and to deal with abiotic stress and competition. However this is not a trivial task. In the following section, an overview is given of the difficulties accounted when aiming at revealing filter-trait relationships.

Chapter 2

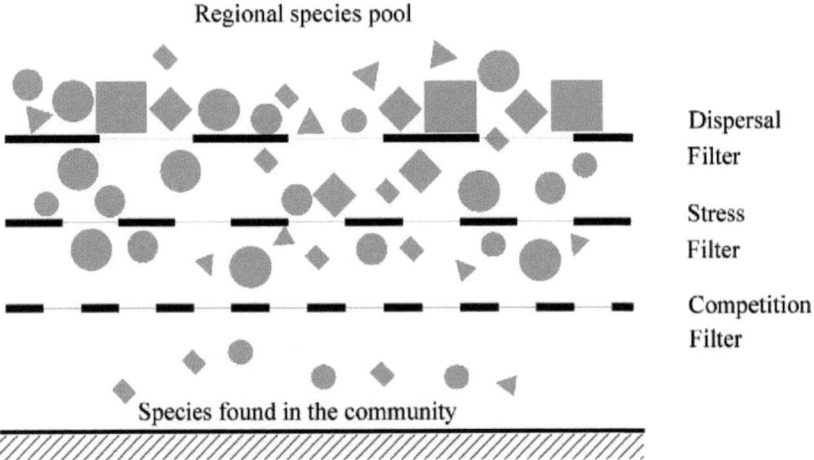

Figure 3: Schematic illustration of community assembly as a hierarchical process. Species from the regional pool of species are able to enter a local community if they (i) manage to arrive at the locality (dispersal filter), find the (ii) local abiotic conditions suitable for recruitment (stress filter), and (iii) are able to withstand the competitive effects exerted by resident species (competition filter). Adapted from Díaz *et al.* (1999).

2.2.2 Assessment of filter-trait relationships

Context-dependency

Filter-trait relationships are highly context-dependent, i.e. different traits may demonstrate important in different situations. For instance, most plants exhibit more than one dispersal strategy. Their seeds may be translocated by wind, water, animals, or other vectors. Dispersal filtering will involve those traits related to the dispersal vector most prevalent in a given landscape. If wind is the prominent dispersal vector, plants should be least dispersal limited if they generate seeds of reduced weight, and eventually possessing uplifting appendices such as a pappus. By contrast, floating properties should play a major role in water-logged systems, whereas hooks and other facilities to attach seeds to fur or feathers should be fostered if zoochory is prevalent (Ozinga *et al.* 2004).

Similarly, the stress filter impacts on those traits which are related to the prominent environmental stressor (Lavorel & Garnier 2002). Empirical studies showed that communities in regularly disturbed habitats were sorted according to their abilities to tolerate disturbance (Kleyer 1999; de Bello *et al.* 2005; Quétier *et al.* 2007): In water-limited ecosystems, stress filtering constrained the exhibition of traits associated with the ability to reduce or avoid water losses (Fonseca *et al.* 2000; Cingolani *et al.* 2007), whereas canopy height, specific leaf area (SLA) and other leaf traits were important along gradients of soil resource availability (Cunningham *et al.* 1999; Cornwell & Ackerly 2009).

Regarding the competition filter, the likelihood to establish and to thrive in the presence of neighbors is a question of resources unconsumed by the extant community (Tilman 1990). Just as the stress filter, competition filtering should therefore impact on traits related to the exploitation of the resources most limited in a given system – with the consequence that stress and competition filter often impact on the same traits. For example, in water limited systems and under strong light competition, characteristics of the root system and the community's canopy structure have been suggested important, respectively (Banta *et al.* 2008). Maximum competition filtering is expected at high densities of species with vast capacities to capture light, water or nutrients. Fargione *et al.* (2003) showed in an experiment that dominance of the species group best adapted to local environmental conditions (C_4-grasses) most successfully inhibited the establishment of colonizing species.

Similarly, Losure *et al.* (2007) observed that tall species substantially reduced colonization rates by lowering the availability of light.

Scale dependency

In addition to their context-dependency, different filter-trait relationships may be observed at different scales. Each filter is most likely to be detected if the scale of observation fits the filter's „grain size" (Bell 2005). This means, to determine the traits that respond to dispersal filtering, communities must be compared that are separated by distances large enough to generate differences in the degree of dispersal limitation. If the considered distances exceed the dispersal range of the majority of species, no functional relationship with dispersal limitation may be detected, simply because all species are limited.

Accordingly, stress filtering is most likely to be detected by consideration of stress gradients and scales that reflect changes in abiotic habitat suitability. Biotic interactions, however, operate between neighboring individuals, and accordingly, evidence for competition arose at fine scales often in the range of a few centimeters (e.g. Wilson & Roxburgh 1994; Wilson & Whittaker 1995; Holdaway & Sparrow 2006).

Community assembly - neutral or trait-driven?

Failure to observe significant filter-trait relationships may not necessarily be attributed to the considered scale and system-specific attributes. Functional traits do not allow to predict the consequences of dispersal, stress or competition filtering if neutral forces decide upon the presence of a species in a community. Considering the vast evidence for trait-driven stress filtering (for reviews see Lavorel & Garnier 2002; Díaz *et al.* 2004), the relevance of traits seems to be most questionable in respect to the dispersal and the competition filter.

Dispersal filtering

Dispersal limitation is often categorized as a neutral process (Hubbell 2001, 2005). There is no doubt, that the probability to migrate from one habitat to another is the complex outcome of many forces, including the timing and the direction of dispersal vectors. Extraordinary strong winds from varying directions may be strong determinants of colonization success eventually overriding differences among species in respect to their typical dispersal distances.

Yet, there is not only empirical evidence that species differ in their abilities to disperse (Clements 1916; Hutchinson 1951), but also that these differences can explain biodiversity patterns in landscapes (Kolb & Diekmann 2005; Ozinga *et al.* 2005; Schurr *et al.* 2007). As a consequence, the neutrality of dispersal filtering during community assembly cannot be assumed but is to be demonstrated.

Competition filtering – a special issue

Contrary to the dispersal filter, the effects of competition on community composition can easily be overlooked because of the effects of the hierarchically higher dispersal and stress filter. This has important implications for the traceability of the competition filter. Already stress filtering will only impact on those species that successfully passed the dispersal filter. But competition filtering can only sort species unaffected by dispersal and stress filtering. Understanding the ways in that dispersal and stress filtering have already constrained the functional diversity of a community is therefore a precondition for the demonstration of competition filtering.

Assembly rules

The obscuring effects of dispersal and stress filtering are, however, only one reason why some researchers have questioned the relevance of competition filtering as a structuring force. According to Hubbell, „... *the number of cases in which local extinction can be definitely attributed to competitive exclusion is vanishing small*" (Hubbell 2001, p. 11). Already Darwin (1859) has cited competition to structure ecological communities. Yet, the general principles behind competition filtering are still subject to considerable controversy. In 1975, Diamond coined the term „assembly rules" to describe the effects of biotic

CHAPTER 2

interactions on community composition - in contrast to the constraints exposed by abiotic stress or dispersal limitation (Box 1). The present work follows Wilson & Gitay (1995) in defining assembly rules as ecological restrictions on the observed patterns of species presence or abundance that are based on the presence or abundance of one or more other species or groups of species. While many assembly rules have been formulated since Diamond (Box 1), two principles are most discussed in a functional context.

The first principle builds on plant geography, sociology and physiology, and stresses the point that species belonging to the same community tend to share similar trait expressions. Principally, such functional similarity may be caused by dispersal or stress filtering, too (see Chapter 6). Yet, it may be competition-induced if dominance of a single superior competitor or group of species leads to the local exclusion of species with contrasting trait expressions (Grubb 1977; Keddy 1992; Grime 2002). On the long run, this process would result in a unimodal distribution of competition-related traits, i.e. in functional convergence (Figure 4) (Navas & Violle 2009).

The second principle, is profoundly influenced by the idea that coexistence in a given place and time requires a minimum degree of niche differentiation (Pianka 1966; MacArthur & Levins 1967). In a functional view, functional similarity is indicative for similar resource use mechanisms. Consequently, competition should be strongest between species with similar trait expression resulting in the exclusion of functionally similar species (MacArthur & Levins 1967; Diamond 1975). The theory of limiting similarity therefore expects an upper limit to the functional similarity of coexisting species creating a bimodal distribution of traits, i.e. functional divergence (Navas & Violle 2009). Those species should be the most successful colonizers that are functionally different to the established species (Diamond 1975).

METACOMMUNITY THEORY AND THE FUNCTIONAL TRAIT CONCEPT

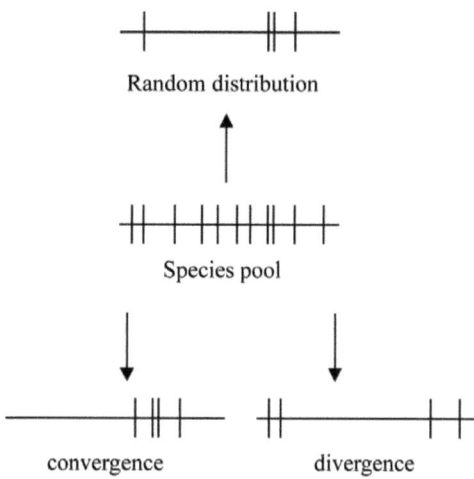

Figure 2: Illustration of the distribution of functional trait expressions following the filtering of traits from a given species pool. Indicated are examples for a random distribution (top), for convergence (bottom left), and for divergence (bottom right).

In summary, ecological theory provides opposing predictions on competition-induced functional patterns (convergence versus divergence), although both build on functional resemblance as the mechanisms behind competitive exclusion. Convergence may be defined as coexisting species being more similar in theirs functional characteristics than expected by chance alone, whereas divergence means that competition results in interacting species being less similar than a random assembly (Figure 4).

The debate about the applicability of these two predictions is still unsolved (e.g. Grime 2006; Wilson 2007). Functional convergence and divergence have received only limited empirical confirmation. This is true for studies searching for convergence and divergence in the composition of established communities (e.g. Watkins & Wilson 2003; Stubbs & Wilson 2004; Thompson et al. 2009), but also for studies investigating competition filtering during the assembly process itself, i.e. when new species attempt to establish in an extant community (e.g. Brown & Kodric-Brown 1979; Janssens et al. 1998; Emery 2007).

Moreover, several studies underlined that colonization success largely depends on the availability of vacant space rather than the functional resemblance between colonizing and resident species (e.g. Burke & Grime 1996; Symstad 2000; Roscher et al. 2009). At a first sight, these observations seem to indicate stochastic colonization processes, questioning the relevance of functional traits during community assembly. The superior role of vacant space was, however, often complemented by the observation, that colonization success was related to the traits of the colonizing species (e.g. Gross & Werner 1982; Tilman 1997; Clarke & Davison 2004). This means that, even if functional resemblance is not important, functional traits might still control colonization success.

2.3 Box 1: Assembly Rules

In 1975, Diamond published his famous work on assembly rules, saying (i) that real communities should contain fewer species combinations than expected by chance and (ii) that the existing combinations should resist invaders. Diamond named competition as the responsible force behind never co-occurring species combinations - or in Diamond's words „forbidden combinations". Forbidden combinations refer to species not sharing similar realized niches although fundamental niches are similar.

Altogether, Diamond formulated ten assembly rules, some of them being based on taxonomic units, others referring to functional relationships. By the while, ecological theory has provided many more hypotheses on the ways in that biotic interactions may influence biodiversity. Wilson (1999) distinguished four categories of assembly rules (Table 1): assembly rules applying for particular species, rules based on presence-absences, rules based on abundance patterns and, finally, rules based on species characteristics.

Diamond's idea of assembly rules is still at the heart of ecological interest. In 2005, the international research project ASSEMBLE (see *http://www.landeco.uni-oldenburg.de/20670.html*) was initiated to investigate the value of trait-based assembly rules for understanding plant community assembly from a functional perspective. The present work was conducted as part of this project with a focus on dispersal processes.

Table 1: Examples for types of assembly rules. Adapted from Wilson (1999).

Type of assembly rule: Rule based on ...	Example	Rule
... particular named species	*Macropygia* assembly rule (Diamond 1975)	*Macropygia ambionensis* and *Macropygia mackinlayi* cannot both be present on any one island.
... presence-absence	Species nestedness (Patterson & Atmar 1986)	Species comprising a depauperate assembly should constitute a proper subset of those in richer assemblies.
... presence-absence	Co-occurrence assembly rule (Diamond 1975)	Some species co-occur less than expected by chance.
... presence-absence	Incidence assembly rule (Diamond 1975)	The incidence and abundance of some species are inversely related to the abundance of other species.
... presence-absence	Checkerboard distributions (Diamond 1975)	Some pairs of species never coexist, either by themselves or as part of a larger combination.
... species abundances	Favored states (Fox 1987)	There is a much higher probability that each species entering a local community will be drawn from a different functional group until each group is represented, before the cycle repeats.
... species abundances	Guild proportionality (Pianka 1980; Wilson 1989)	Different guilds are represented in a relatively constant from arising from the hypothesized situation that competition between two species is stronger when they are both in the same guild.

Table 1: Extended.

Type of assembly rule: Rule based on ...	Example	Rule
... characters	Limiting similarity (MacArthur & Levins 1967)	There is a limit to how similar species can be in their resource requirements and still co-exits – if a pair of species are too similar, one will be competitively excluded.
... characters	Constant body size ratios (Brown 1973; Dayan & Simberloff 1994)	Species of similar body size coexist less frequently in local communities and overlap less in geographic distributions than is expected on the basis of chance.
... characters	Texture convergence (Schluter 1990; Wiens 1991)	Ecomorphologically (functionally) similar species replace each other in structurally similar habitats although the identity of the species occupying a particular niche may change.

Conclusion

The functional group approach allows generalizations as it predicts biodiversity response at a trait base rather than a species base. In a functional framework, community assembly may be understood as a series of hierarchical filters. From the total species pool, those species are excluded whose functional attributes do not fit the local requirements for dispersal as well as stress and competition tolerance. Hence, the definition of functional groups requires knowledge of the relationships between filters and functional traits.

However, functional relationships with the dispersal, stress and competition filters are not generally established. The relevance of a given trait is highly context and scale dependent making it impossible to determine traits in advance. Moreover, functional relationships with the dispersal and the competition filter are not generally supported, i.e. these processes could be trait neutral as well. Evidence for competition filtering may have been hampered for two reasons. Firstly, competition induced functional patterns may be obscured by hierarchical higher filters (dispersal and stress filter). Secondly, ecological theory has provided contrasting predictions on the effects of competition filtering on functional community composition. In summary, the determination of ecologically meaningful plant functional groups in any given ecosystem requires

- to identify for each filter the relevant traits at adequate scales,
- to take into account the obscuring effects of hierarchically higher filters, and
- to regard alternative hypotheses of functional trait distributions within a community.

2.4 Reconciling metacommunity ecology and the functional trait concept

The functional group approach may provide a valuable tool for the validation of metacommunity paradigms in several respects. Consideration of functional groups rather than species identities narrows the number of responsive variables during community assembly, thus reducing the complexity of metacommunity research. Moreover, the observation of functional relationships with the dispersal, stress and competition filter allows conclusions on the applicability of metacommunity paradigms for the studied communities (Table 2).

If species do not differ in their responses to spatial habitat configuration, environmental stress and competition neutral processes should prevail. General absence of functional relationships is consistent with ecological equality of species highlighting the relevance of the neutral metacommunity paradigm. A patch dynamic view is supported if stress filtering is negligible but dispersal and competition filtering are sustained, and there is a trade-off between traits associated with dispersal and competitive abilities. By contrast, stress filtering points to the species sorting or the mass effects paradigm. Further functional relationships are not required but may be observed. For instance, the level of dispersal is an important criterion to distinguish the species sorting and the mass paradigm, but no differences among species dispersal capacities are required. However, if a functional relationship with dispersal filtering indicates that species vary in their dispersal capacities, the mass effects paradigm may apply to good dispersers but the species sorting paradigm to those with low dispersal capacities (Leibold *et al.* 2004).

CHAPTER 2

Table 2: Decision tree for the metacommunity paradigm expected from the observation or non-observation of relationships between functional traits and dispersal, stress and competition filtering. For each filter, fills denote no relationship („-"), confirmed relationship („x"), and a relationship that may or may not be observed („(x)").

Dispersal filtering	Stress filtering	Competition filtering	Expected paradigm
x	-	x	Patch dynamics paradigm
-	-	-	Neutral paradigm
(x) *	x	(x)	Mass effects paradigm
(x) **	x	(x)	Species sorting paradigm

* high dispersal ability
** low dispersal ability

Aims

Integration of functional ecology and metacommunity theory may decidedly promote our understanding of the mechanisms that generate and maintain biodiversity patterns in a landscape. In this work, a functional group approach was employed to validate metacommunity mechanisms in a real landscape. Relevance of the four metacommunity paradigms was derived from relationships between functional traits and the dispersal, the stress and the competition filter. Functional relationships with the stress gradient were interpreted as evidence for the species sorting or the mass effects paradigm, whereas functional dispersal filtering was in favor of the patch dynamics paradigm. Absence of functional relationships was understood as support of the neutral paradigm. The aim of this work was to answer the following questions:

- What are the effects of the dispersal filter, the stress filter and the competition filter on the functional diversity of local plant communities and the constituent metacommunity?

    - What traits are responsive to what filter?

    - What filters are trait neutral?

- What plant functional groups constitute the metacommunity?

- What may be concluded on the relevance of the four metacommunity paradigms for governing biodiversity in the study area?

"In questions of science, the authority of a thousand is not worth the humble reasoning of a single individual."

Galileo Galilei

Chapter 3

3 Study design

3.1 Study area

The exploration of metacommunity dynamics in a functional group framework may be promoted by two characteristics of the study area. Firstly, the area should exhibit an isolation gradient along which plant functional groups with different dispersal abilities may separate. The likelihood of observing groups with low dispersal abilities should decrease with increasing spatiotemporal isolation of a given habitat. Secondly, the study of functional trait distributions requires increased species numbers with associated variability in traits.

Urban landscapes meet these conditions in unique ways. They comprise a patchwork of naturally developed brownfields or wastelands interspersed between paved structures and intensively managed vegetation types (e.g. lawns). Depending on the geographic distances separating brownfield patches, inhabiting plants face varying degrees of spatial isolation with consequences for the exchange of individuals between local populations or the colonization of unoccupied patches. The term "patch" is used here to denote a discrete area with homogeneous habitat conditions (suitable or unsuitable).

Due to the highly variable life span of brownfields, urban landscapes may be considered as very dynamic ecosystems (Muratet *et al.* 2007). During urban development, brownfields are typically replaced by buildings or parks, while in other locations industrially used areas are abandoned and new brownfields emerge (Kattwinkel *et al.* 2009). From a plant perspective, this turnover implies the regular deterministic extinction of local populations, but also the chance to compensate for these losses by colonization of new habitat patches. Consequently, the survival of plants in dynamic landscapes is conditioned on dispersal among habitat patches, which provides an excellent setting for exploring the contribution of dispersal to the generation of biodiversity patterns.

The abandonment of formerly used areas is also the starting point for the succession of plant communities. During succession more and more species overcome dispersal limitation and colonize, which is at least partly responsible for compositional changes between successional stages (Cook *et al.* 2005). Accordingly, successional age may be assumed a proxy for temporal habitat isolation, adding a temporal dimension to the gradient of spatial habitat isolation in urban landscapes.

Urban brownfields are typically rich in ruderal plant species, and particularly, when they are still young (Godefroid & Koedam 2007; Schadek *et al.* 2009). In a study in the Greater Paris Area, urban brownfields comprised 58% of the biodiversity of the region with maximum species richness in brownfields with an age between 4 and 13 years (Muratet *et al.* 2007). Similarly, Angold *et al.* (2006) reported highest diversity in brownfields younger than 20 years. Kleyer (2002) and Schadek *et al.* (2009) have demonstrated the usefulness of the functional group approach in urban brownfield to describe community changes along gradients of disturbance, resource availability and successional age.

The present work was conducted in a newly developed industrial park which, due to its extraordinary historical record, combined spatial and temporal gradients of isolation in a unique way. The study area was situated at Bremen, a city in North-West Germany (53°05′N, 8°42′E, 2 m a.s.l.) with a temperate, maritime climate (mean annual temperature: 8.8°C, mean annual precipitation: 753 mm, period 1961 – 1990, Deutscher Wetterdienst 2006/07). It has been created in a stepwise procedure over 40 years. Starting in the early 1970s, increasing fractions of the area have been filled in with about 2 m of non-autochthonous sandy sediments (Figure 5). By 2006, the area sustained approximately 4 km^2 of isolated sandy habitats surrounded by moist marshland.

STUDY DESIGN

Figure 3: Sequence of aerial photographs of a subarea of the study area from 1974, 1987 and 2002.

While large fractions of the created land have been built up, ruderal plant communities developed in the remaining areas. In this work, these remaining areas with naturally developed plant communities are denoted as "brownfields" although they do not conform to another common understanding of brownfields: as derelict or abandoned land that is no longer used. By contrast, the brownfields of the study area have never been used for commercial purposes but are reserved for future industrial development.

The brownfields spanned an age gradient of almost 40 years with associated differences in the degree of successional progress. Accordingly, this age gradient was regarded as a proxy for the degree of temporal isolation. The gradient of spatial isolation was provided by the patchy nature of brownfields which formed a network of potential habitats separated by a matrix of impervious areas. This matrix was composed of buildings, streets and parking lots uninhabitable for plants. Thus, the likelihood of missing plant occurrences in the matrix (matrix effect) was minimal.

Within-patch disturbances (i.e. local damages to established vegetation) were frequently encountered in the study area. Out of 95 plots, 44 plots (46%) were subject to some type of disturbance between 2006 and 2008. Vegetation damages were most often caused by the passage of heavy vehicles or the periodic deposition of construction material. Associated damages to the vegetation sward did however not eliminate the chance of regeneration from the seed bank, i.e. succession after disturbance included dispersal in time. By contrast, in newly created brownfields, the course of succession depended on spatial dispersal processes as no local seed bank was available (see Chapter 5). Since sandy filling materials were used consistently during development of the study area, the area was expected to be rather homogeneous in respect to edaphic conditions.

3.2 Sampling design: Bridging scales

To reveal the effects of dispersal, stress and competition filtering, biodiversity patterns were studies at three scales (see Figure 6). The dispersal filter was investigated at the patch scale (Figure 7). Thereto, the entire study area was subdivided into 958 approximately equal-sized patches and incidence information collected for each patch (see Chapter 4 for more details).

STUDY DESIGN

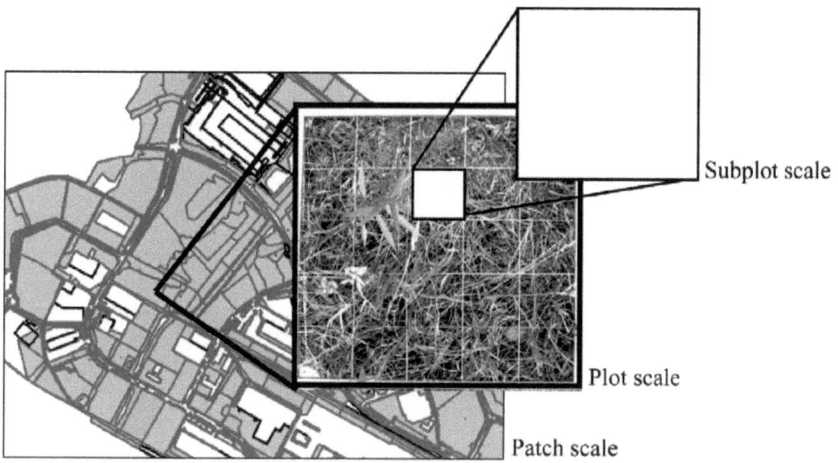

Figure 6: Overview of observation scales considered to assess the effects of dispersal (patch scale), stress (plot scale) and competition filtering (subplot scale).

Stress filtering was studied at the plot scale, i.e. in a number of representative 1 m² plots. These plots were located within brownfield patches according to a stratified sampling design. The corresponding spheres were provided by brownfield age assuming that age can be used as a proxy for temporal habitat isolation. In each plot, frequency information for all vascular plants was recorded by counting the occurrences of each species in 100 subplots. Each subplot measured 10 x 10 cm, one hundred of them adding up to one plot.

The impact of competition filtering was assessed at the subplot scale. To picture biotic interactions, including the effects of shading, the above ground projection was considered rather than the mere rooting point of each plant.

CHAPTER 3

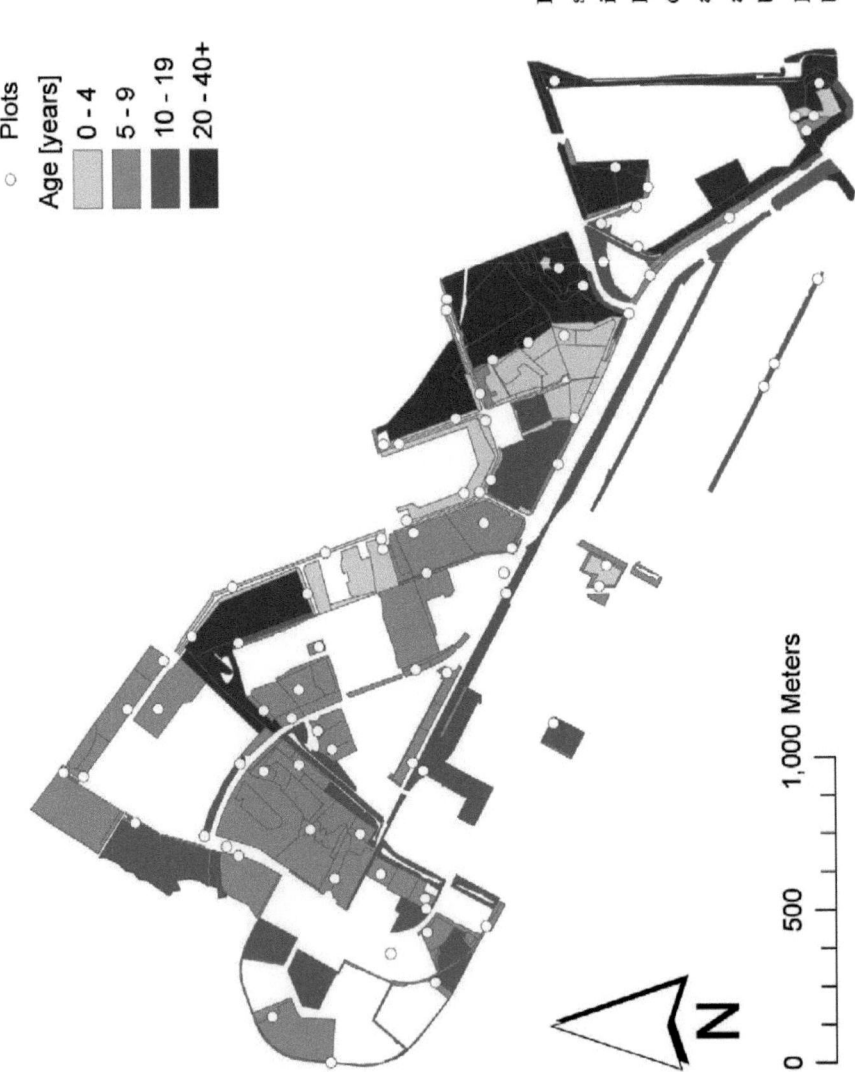

Figure 4: Overview of the study site, the subdivision into patches and the location of plots (white dots). Colors indicate the approximate year of creation of patches inhabitable by plants. White areas represent matrix (paved or built up areas, marshland).

STUDY DESIGN

Vegetation data at patch and plot scale was complemented by scale-specific descriptors of environmental stress. At the patch scale, stress was assessed by averaged Ellenberg values, whereas soil measurements were interpreted at the plot scale (see Chapter 4 and Chapter 5 for more details). For logistic reasons, environmental differences were not explored at the subplot scale.

There is no general consensus on how to disentangle the effects of overlapping filters. Therefore, for each analysis the methodology was applied that seemed to be most appropriate to account for the obscuring effects of hierarchically higher filters on stress or competition filtering. In two analyses, statistical methods served that purpose. Species distribution models (Chapter 4) as well as constrained null models (Chapter 6) account for dispersal and stress filtering by removing the variation explained by spatial and environmental predictors. The remaining variance may then be interpreted in terms of competition filtering (Legendre & Legendre 2006). A third analysis (Chapter 5) took advantage of seed bank information. Seed bank analysis avoids the problem of dispersal limitation because seed banks comprise only those species that have definitely been able to arrive at a given locality. Discrepancies between seed bank and vegetation composition may thus be interpreted in respect to stress and competition filtering.

3.3 Functional diversity

3.3.1 Trait selection

Functional diversity was assessed by "soft traits" rather than "hard traits" (Hodgson *et al.* 1999; Weiher *et al.* 1999). Hard traits are directly linked to central functions of plant life (e.g. growth, reproduction, survival rates), but require laborious and/or expensive measurements. By contrast, soft traits are undemanding or easily-measurable morpho-physiological plant features which may be used as surrogates of these functions (Violle *et al.* 2007). For instance, specific leaf area (SLA) exhibits close relationships with relative growth rate (Garnier *et al.* 1997) and net photosynthetic rate (Wright *et al.* 2004b).

There are two kinds of soft traits: those that are associated with the response or sensitivity of a plant to changes in environmental conditions (response traits), and those that describe the effect or impact of a species on its environment (effect traits) (Goldberg 1996; Violle *et al.* 2009). Distinguishing between response and effect traits is particularly important in respect

CHAPTER 3

to competition. Different traits may lead to different rankings of species in terms of competitive response and effect (Goldberg & Barton 1992). In this work, the main interest was on traits that enable the estimation of:

- plant response to dispersal filtering
- plant response to stress filtering
- plant response to competition filtering
- competitive effects exerted on neighboring plants

The main challenge for plant functional group classification lies in choosing which attributes to include since an element of subjectivity is unavoidable (Gitay & Noble 1997). The logical consequence is to group species *a posteriori*, i.e. responsive traits have to be determined before differentiating ecological strategies (for examples see Díaz *et al.* 1999). This approach was followed in Chapter 4.

Yet, this methodology is not always convenient. Particularly, the assessment of traits associated with competitive response and effect requires an experimental investigation of pairwise interactions, which is hardly feasible for large species sets – and is then still not comparable to field conditions. Therefore, another approach was followed in Chapter 6. For each of the listed four types of traits or functions of interest (Table 3), candidate functional traits were selected (i) according to their relevance for these functions following literature and (ii) according to their availability in trait databases. Based on these candidate traits, observed functional patterns in real communities could be compared to random expectations allowing conclusions on operating filters.

A third approach consists in testing filter processes on hypothesized functional groups. This is a standard approach in science and is justified as long as the considered functional traits are related to the investigated function of interest (Wilson 1999). It may be regarded superior to the previously described approaches, as it reduces the variety of possible trait combinations to a handy number of combinations realized in a given set of species. Striving for maximum comparability to natural conditions, such an approach was followed in Chapter 5, when exploring the role of functional resemblance during community assembly.

STUDY DESIGN

Table 3: Overview of functional traits considered in this work, and their predicted relevance to describe (i) plant response to dispersal filtering, (ii) plant response to environmental filtering, (iii) plant response to competition filtering, and (iv) the competitive effect of plants exerted on neighbors (see references for details). "High" (or "low") indicates that responsiveness or effect increases (or decreases) with increasing values of the trait.

Functional trait	Response to			Competitive effect
	dispersal	stress	competition	
Seed number	Low [1]			
Terminal velocity	High [2, 3]			
Seed mass	High [4, 5]		Low [14, 15]	
Seed bank longevity	Low [6]	Low [10, 11]		
Specific leaf area (SLA)		High [5, 7, 8]	Low [15]	High [19, 20]
Life span		High [9, 10]		High [10, 15]
Lateral spread/clonal growth		Low [12, 13]	Low [14]	High [5, 17, 18]
Canopy height			Low [16]	High [5, 17, 18]
Flowering month				High [5]
Leaf size				High [21]

[1] Kolb & Diekmann (2005); [2] Tackenberg et al. (2003); [3] Grashof-Bokdam & Geertsema(1998); [4] Grubb (1987); [5] Grime (2002); [6] Ozinga et al. (2005); [7] Cunningham et al. (1999); [8] Fonseca et al. (2000); [9] Bossuyt & Honnay (2006); [10] Eriksson (1996); [11] Warner & Chesson (1985); [12] Reynolds et al. (2007); [13] Baer et al. (2004); [14] Goldberg & Barton (1992); [15] Weiher et al. *(1999)*; [16] Crawley et al. (1996); [17] Anten & Hirose (1999); [18] Gross et al. (2007); [19] Garnier (1992); [20] Fonseca et al. (2000); [21] Hodgson et al. (1999)

CHAPTER 3

3.3.2 Trait data

Locally collected trait information is often regarded superior to pooled measurements obtainable from trait-databases. Thanks to a previous study in the same area (Schadek 2006), local functional information for the most prevalent species was largely available and required little complementation. Yet, for logistic reasons, it was impossible to collect data on the entire species pool (255 species, see Appendix 3). Therefore, local data was used when considering exclusively the most prevalent plant species (Chapter 4) but local data and trait information from the LEDA trait base (Kleyer *et al.* 2008) were combined to assess the functional diversity of whole communities (Chapter 5 and Chapter 6).

3.3.3 Assessment of functional resemblance

Assessment of the role of functional resemblance for competition filtering requires a measure of the functional resemblance for a given species set. Ecological theory knows a vast number of functional diversity measures but most lack important features. More specifically, such a measure should allow to consider multiple traits, to deal with mixed data types, and to compare local communities (Laliberté & Legendre 2010).

This work followed Villéger *et al.* (2008) in assuming functional diversity a n-dimensional functional space in which each trait provides one dimension. The functional diversity of a given local community can be described by the distribution of traits of species forming that community in the functional space span by the reference species pool. Villéger *et al.* (2008) proposed three measures:

- Functional richness
- Functional evenness
- Functional divergence

These three measures have been extended by functional dispersion as a fourth measure (Laliberté & Legendre 2010). Functional dispersion may be regarded as an equivalent of functional divergence, but additionally incorporates information on functional richness. Contrary to functional convergence, dispersion captures an increase in dissimilarity with increasing richness which may be important when comparing the functional resemblance of communities occupying different trait volumes (Figure 8). As a consequence, all four

STUDY DESIGN

measures (functional richness, evenness, divergence and dispersion) were considered for the analysis of functional community composition in Chapter 6.

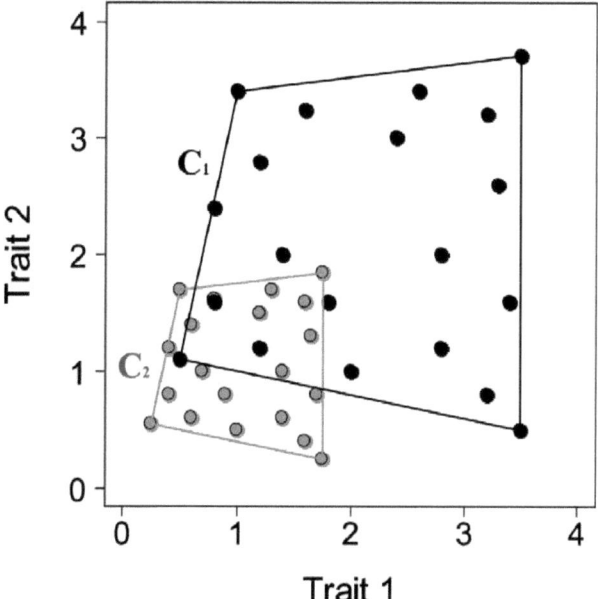

Figure 8: Comparison of functional divergence and functional convergence. Two communities, C_1 (grey circles, grey polygon) and C_2 (black circles, black polygon), each composed of 20 species. Although the species in C_1 are obviously less dispersed in two-dimensional functional trait space than the species in C_2, both communities obtain a functional divergence value of 0.808, In contrast, the functional dispersion values of C_1 and C_2 are 0.697 and 1.395, respectively. By Laliberté & Legendre (2010).

Chapter 4

"Like the end of the rainbow, the tail of the seed dispersal curve (...) is impossible to reach. The occasional seed is carried quite extraordinary distances by chance events, but these seeds are so few that we can only ever know where they end up when they attract attention by starting a new population in an alien site."

Jonathan W. Silvertown

4 Seed number and terminal velocity determine plant response to habitat connectivity in an urban landscape

Andrea Schleicher

Robert Biedermann

Michael Kleyer

Summary

Identification of trait syndromes that make species vulnerable to habitat fragmentation is essential in predicting biodiversity change. Plants are considered particularly vulnerable if their capacities for persistence in and for dispersal among local habitats are low. Here we investigated the influence of easily measured functional traits on the presence of 45 plant species in an urban landscape in north-west Germany where patches were separated by distances consistent with regular plant dispersal range. To describe the spatial configuration of patches we calculated species-specific patch connectivities. Then we assessed plant connectivity responses in distribution models calculated from connectivities and environmental predictors. Twenty (45%) of the analysed species showed a positive connectivity response after accounting for species-specific habitat requirements. These species differed from non-responsive species by functional traits associated with dispersal, including reduced seed numbers and higher terminal velocities relative to non-responsive species. Local persistence traits played, however, no role, which we attribute to the environmental conditions of urban habitats and their spatiotemporal characteristics. Our study underlines that even ruderal plants experience dispersal limitation, and demonstrates that easily measured functional traits may be used as indicators of fragmentation vulnerability in urban systems allowing generalizations to larger species sets.

CHAPTER 4

4.1 Introduction

Habitat fragmentation is considered as one of the main factors leading to loss of biodiversity. When habitats become increasingly dissected, species are confronted with habitats of reduced size, extended isolation and novel ecological boundaries with major consequences for their regional persistence (Fahrig 2003; Ewers & Didham 2006). However, functional differences among species may result in different responses to fragmentation. Therefore, identification of species trait syndromes with increased vulnerability to fragmentation is pivotal in predicting biodiversity in fragmented landscapes (Henle et al. 2004; Ozinga et al. 2009).

Response to fragmentation depends on local extinction risk and a species potential to rescue declining populations via (re-) colonization (Vos et al. 2001; Johst et al. 2002). Plants will be most vulnerable to fragmentation if their capacities for local persistence and dispersal among patches are low. Translated to a functional perspective, this means that traits enabling a species to survive within and to move between habitat patches confer the ability to maintain viable (meta-) populations in fragmented landscapes (Johst et al. 2002; Verheyen et al. 2004).

The ability of plants to colonize suitable habitats depends on both dispersal range and the number of potential dispersers (Johst et al. 2002; Nathan 2006). This underlines the functional relevance of two types of plant traits. First, dispersal range is determined by traits such as terminal velocity or floating capacity of fruits, which have been used to describe the species-specific component of seed dispersal kernels (e.g. Tackenberg et al. 2003). Second, the number of seeds a plant can generate indicates the number of potential dispersers. The production of many seeds may increase the chance of colonization even if the dispersal range is limited (Grashof-Bokdam & Geertsema 1998).

By contrast, relationships between local persistence ability and plant functional traits are widely recognized as highly ecosystem dependent. Under ample resource conditions, species are favored that are able to grow and reproduce in the presence of competitors (Goldberg & Barton 1992; Suding et al. 2003). In stressful environments, however, it is important to cope with physically low resource levels. Consequently, a shift from traits conferring high competitive ability towards stress tolerance traits occurs along a gradient of decreasing productivity. Shifts in leaf and stem trait expressions are expected, e.g. from high to low specific leaf area, canopy height and leaf size (Díaz et al. 2004). Increasing stress or disturbance may further select trait expressions that enable plants to bridge periods of unsuitable habitat conditions (e.g. persistent seed banks) or rapidly regenerate between disturbance events (e.g. short life span).

In this study, we investigated the role of plant functional traits in response to habitat fragmentation in an urban environment. We assumed that traits supporting local persistence and dispersal ability influence plant response to fragmentation. Specifically, we hypothesized that species with distributions negatively affected by fragmentation will display trait expressions that indicate the following: (i) decreased dispersal range and lower dispersal probability; (ii) decreased capacity to persist locally via seed banks; and (iii) decreased ability to tolerate stress, disturbance or competition relative to species less sensitive to the effects of fragmentation.

Previous studies have documented not only vast discrepancies regarding the commonness of fragmentation effects. Within large species data sets, responsiveness to fragmentation ranged between rare responses (e.g. Dupré & Ehrlén 2002; Krauss et al. 2004) to connectivity being the main predictor (Bastin & Thomas 1999; Piessens et al. 2005). Studies have also reported a variety of relationships between fragmentation vulnerability and functional traits, which seems to exceed the expected variability between ecosystems. For instance, although Kolb & Diekmann (2005) and Dupré & Ehrlén (2002) conducted their studies in deciduous forests with similar fragmentation characteristics, Kolb & Diekmann identified many more traits responding to fragmentation, including canopy height and seed longevity. These discrepancies may to some extent be explained by differences in methodology (Freckleton & Watkinson 2002; Tremlová & Münzbergová 2007).

The relevance of dispersal limitation for biodiversity patterns is highly dependent on the study scale (Gilbert & Lechowicz 2004; Girdler & Barrie 2008). For instance, field and modeling studies have demonstrated that dispersal by wind rarely covers distances of more than 100 m, whereas dispersal across larger distances becomes increasingly dependent on stochastic factors (Higgins et al. 2003; Tackenberg et al. 2003; Soons et al. 2004). Regarding that much previous work has focused on habitat patches separated by several kilometers (e.g. Dupré & Ehrlén 2002; Kolb & Diekmann 2005; Piessens et al. 2005) the

CHAPTER 4

rarity and unpredictability of dispersal events is likely to have hampered the detection of relevant functional characteristics. In addition, incorrect assumptions on matrix uninhabitability and habitat suitability may obscure the influence of fragmentation. Discriminating suitable and unsuitable habitats is particularly problematic in plants that do not have easily identified niches (Freckleton & Watkinson 2002; Ehrlén *et al.* 2006). Consequently, habitat suitability should be estimated for each species separately, rather than assuming identical niche requirements for species inhabiting a given habitat type. For instance, regarding all forest fragments suitable habitat for understory species is error-prone as subtle differences in e.g. soil acidity may suffice to make a habitat uninhabitable for a subset (e.g. Dupré & Ehrlén 2002). Ruling out undetected incidences in the matrix is facilitated in urban environments. These are particularly suited for fragmentation studies because impervious surfaces represent an uninhabitable matrix that is clearly distinguished from potential habitats. From a statistical point of view, the effects of spatial autocorrelation (SAuC) should be considered when distinguishing unsuitable from uncolonized habitat. SAuC is well-known to result in erroneous assessment of significant explanatory variables violating the assumption of independent samples (Legendre & Legendre 2006; Dormann *et al.* 2007). However it becomes more than a statistical issue in empirical studies regarding dispersal limitation. Spatially clustered plant species occurrences are not necessarily generated by dispersal but may simply reflect spatially autocorrelated suitable habitats (Legendre & Legendre 2006). Therefore, the effects of dispersal limitations must be separated from the effects of spatially autocorrelated environmental conditions.

In this study, we explicitly aimed to overcome the obscuring effects arising from the above-mentioned methodological challenges. Therefore, we investigated the relationship between plant functional traits and fragmentation vulnerability at a scale that corresponds to the dispersal range most commonly observed for anemochorous plants. We predicted that clear differences among species would emerge in an urban landscape in which impervious surfaces help to rule out matrix effects, and when the spatial analysis of patch configuration is based on the specific habitat requirements of each single species. Additionally, we developed a methodology that separates dispersal limitation and habitat unsuitability, and accounts for the effects of spatial autocorrelation. This approach enabled us to identify easily measured functional traits corresponding to fragmentation vulnerability, which may be tested and subsequently applied to different groups of taxa.

4.2 Methods

4.2.1 Study area

The study was conducted at an industrial park at the city of Bremen, Germany (53°05´N, 8°44´E, mean annual temperature 8.8°C, mean annual precipitation 694 mm: Deutscher Wetterdienst 2006/07) which constituted an artificial island habitat. The area had been created by filling the original marshland with about 2 m of sand in a step-by-step procedure starting in the 1970s. By 2006, this process had generated a patch network of about 4.0 km² and an age gradient spanning almost 40 years.

In determining habitat connectivity, the likelihood of missing species occurrences in the surrounding landscape composed of moist grassland was small due to the strong contrast with the land-fills in terms of environmental conditions and species composition. Moreover, buildings and paved structures clearly divided the site into an uninhabitable matrix of impervious surfaces without any plant occurrences, and habitable patches which comprised wastelands, road verges and railroad tracks dominated by ruderal plant communities. Patch geometries were mapped in a geographic information system (GIS), whereby patches larger than 100 x 100 m were split into smaller equally sized patches.

Between May and September 2006, we visited all resulting 958 patches and collected presence/absence data of 52 plant species. Species selection aimed at satisfying the statistical requirement of a minimum of ten occurrences which has been advised for logistic regression analysis (Steyerberg *et al.* 2001). That is, we collected data on those species which had been observed with a minimum occurrence frequency of ten in a previous study (Schadek *et al.* 2009). As an inverse measure of isolation, we determined species-specific connectivity for each patch as proposed by Hanski (1994). The connectivity S of patch i to all patches j was calculated as follows:

$$S_i = \sum_{j \neq i} \exp(-\alpha * d_{ij}) * A_j^b$$

where A_j is the area of patch j, d_{ij} is the distance between the centroids of patches i and j, and $b=0.5$ following the recommendations of Moilanen & Nieminen (2002). We used $\alpha=0.02$, which corresponds to a migration distance of 100 m as a realistic value for intermediate and frequently observed dispersal range of many anemochorous plants (Tackenberg et al. 2003; Soons et al. 2004). Furthermore, Moilanen & Nieminen (2002) reported that connectivity is not highly sensitive to the value of α. Centre to centre distances between patches and patch areas were obtained using GIS (ESRI Inc. 2006).

To disentangle suitable from unsuitable habitats for each species we calculated species distribution models with the following variables: soil texture (paved, coarse-grained, fine-grained), degree of shrub encroachment, typical disturbance regime and patch age (age since soil deposition) based on chronosequences of aerial photos and field observations. Due to the large number we could not directly record soil properties for each patch (n=958). Instead, we calculated average Ellenberg values for soil reaction and moisture to describe community position along a productivity gradient. Ellenberg values associated with soil pH and water availability were chosen as they had proven to be most suitable to describe community composition in the study area (Schadek et al. 2009). The average Ellenberg value of a patch was calculated from the values of the species present in the patch as obtained from Ellenberg (1986). Modeled species were omitted from the calculation to avoid circularity. Ellenberg values show a close correlation with corresponding measurements of environmental variables (Schaffers & Sykora 2000) and have proven suitable to assess local environmental habitat conditions (Ozinga et al. 2004; Verheyen et al. 2004). Furthermore, the obtained values conformed well to the spatial patterns suggested by the measurements of Schadek et al. (2009).

4.2.2 Functional trait data

Because wind was assumed to be the most prevalent dispersal vector in the urban landscape, seed terminal velocity was chosen as a species-specific measure of dispersal range. Such a simple approach is adequate in herbaceous vegetation where stochastic factors such as wind turbulences play a minor role (Fenner & Thompson 2004). Seed releasing height was not used because this trait was strongly correlated with canopy height (see below). The high number of patches made it logistically impossible to determine seed production per area. Therefore, seed number per species was used to estimate dispersal probability.

Species ability to persist locally was evaluated by two sets of traits. Local persistence under disturbance was described by life span and seed bank longevity index (Bekker et al. 1998) the latter quantifying the capacity of a species to maintain a population at a site from the available seed bank. Ability to persist in either resource-poor or resource-rich environments was assessed by specific leaf area, seed mass, leaf size and canopy height (Westoby et al. 2002). Local trait information had been collected in a previous study in the area (Schadek 2006) with the exception of plant life span and seed longevity index which were extracted from the LEDA Traitbase (Kleyer et al. 2008). All traits were collected according to the LEDA Traitbase standards (see www.leda-traitbase.org). Species nomenclature followed Jäger & Werner (2002).

4.2.3 Statistical analysis

Statistical analyses were performed on a subset of 268 patches to exclude communities subject to regular fertilization and maintenance management (e.g. lawns, parkings, etc.). However, connectivities of the species in the 268 patches were calculated from occupancy data of all 958 patches. Moreover, the reduction did not imply the elimination of any disturbance effects in the dataset. In nearly 20% of the remaining patches the vegetation was visibly damaged due to nearby construction activities.

To determine the response of each species to connectivity we combined a habitat modeling approach with a methodology to handle spatial autocorrelation. Our approach involved altogether five steps which can be summarized as follows: We first determined the response of species to environmental variables and connectivity following Strauss & Biedermann (2006, steps 1 and 2). Then, we performed two tests aiming to account for the effects of spatial autocorrelation. In a first test, we addressed the problem of biased variable estimates arising from spatial data dependency (step3). The second test specifically aimed at ensuring that our measure of connectivity did not simply reflect autocorrelated environmental conditions (step 4). In a last step, we applied model averaging following Burnham & Anderson (2002) to merge adequate habitat models into one averaged model per species. The single steps will be described in the following in more detail.

As a first step, we calculated univariate logistic regression models for each species. Thereby, we tested for either sigmoid or unimodal relationships between species occurrence, abiotic patch parameters and species-specific connectivity (Strauss & Biedermann 2006). Patch age and connectivities were log-transformed before use in the logistic regressions.

CHAPTER 4

In the second step, we calculated multiple models for all possible combinations of the significant variables identified from the univariate models to avoid spurious inclusion of variables (Strauss & Biedermann 2006). Multiple models were only chosen for further analyses if the following criteria were met: (i) the model performed better than any model of a lower hierarchy; and (ii) if coefficients significantly differed from zero ($p < 0.05$).

As a third step, we corrected for biased parameter estimates resulting from spatial data dependency by capturing the spatial patch configuration in spatial vectors and adding them to the model (Dray *et al.* 2006; Dormann *et al.* 2007). We computed global Moran's tests to determine if spatial structure was present in the model residuals. Second-order stationarity assumption was verified by visually checking variograms for a sill and additionally applying Levene's test for variance equality on three different patch groupings (Legendre & Legendre 2006): (i) based on increasing x-coordinates; (ii) based on increasing y-coordinates; and (iii) based on boundaries as imposed by road and railway infrastructure. If the second-order stationarity assumption was not accepted, we considered models with a p-value >0.01 of any coefficient likely to be erroneously significant and treated them like models with residuals exhibiting a significant spatial structure according to Moran's test. If spatial structure was detected in the residuals, we applied spatial eigenvector mapping (Dray 2006). Spatial eigenvector mapping (SEVM) is particularly suitable to check variable significance because it removes residual spatial autocorrelation and can deal with different error distributions such as that originating from binary response data (Griffith & Peres-Neto 2006; Dormann *et al.* 2007). We followed Dormann *et al.* (2007) and calculated spatial eigenvectors which were then incorporated into the regression model as covariates. If any model coefficient became insignificant after the addition of the SEVM vectors the model was excluded from further analysis.

To further separate dispersal effects from those of spatially autocorrelated environmental factors we applied an additional test to models exhibiting connectivity (step 4). We assumed that connectivity captured not only spatial patterns generated by autocorrelated environmental factors but also patterns induced by dispersal processes between local population (Legendre & Legendre 2006; Dormann *et al.* 2007). If so, connectivity should explain species occurrences better than the SEVM vectors which are based solely on the spatial positions of the habitats. Hence, we calculated a corresponding model in that SEVM vectors replaced connectivity. If this model outperformed the connectivity model in a likelihood ratio test we concluded that connectivity reflected spatially autocorrelated environmental factors and excluded the model from further analysis.

In a fifth step, we applied model averaging as proposed by Burnham & Anderson (2002) to all remaining multiple models for a species. Model averaging does not rely on a model selection procedure, but rather merges all entering models to a single averaged model. Each entering model is assigned a weight according to its relative performance calculated as the difference between the Akaike Information Criterion (AIC) of the entering model and the AIC of the best model. In addition to averaged coefficients, model averaging calculates a variable weight according to variable performance and number of models containing that variable. This variable weight can be used to assess the importance of a predictor relative to other predictors (Burnham & Anderson 2002) and we used connectivity weight to determine species connectivity relevance in comparison to the importance of environmental factors. Moreover, we defined two connectivity response groups: Species with a connectivity weight smaller than one percent were classified as non-responsive to connectivity, and species with a connectivity weight greater than one per cent were classified as responsive to connectivity (see Table 6).

Trait relationships were analyzed using Spearman correlation coefficients or Kolmogorov-Smirnov tests and differences between connectivity response groups with Wilcoxon-Mann-Whitney and Chi-Square tests. Seed mass and seed number were log-transformed prior to analysis. Interactions between responsive functional traits were calculated by generating the products or quotients and comparing the obtained values between the connectivity response groups. Species models were calculated using the statistical programming software R (R Development Core Team 2005) and other statistical analyses were performed in SPSS 15.0 for Windows (SPSS Inc. 2006).

4.3 Results

4.3.1 Importance of habitat isolation versus habitat suitability

For 45 out of 52 species our analysis yielded averaged models with Nagel-kerke's $R^2 > 0.1$. All models with Nagelkerke's $R^2 < 0.1$ were omitted from subsequent analyses. Within the remaining models Nagelkerke's R^2 ranged between 0.11 and 0.33 (mean: 0.20, Table 6).

CHAPTER 4

Model averaging provides variable weights that can be used to assess variable importance (Burnham & Anderson 2002). Accordingly, we derived a given species connectivity response from the weight of the connectivity variable in the species model. Out of 45 remaining species, 25 species (56%) possessed connectivity weights >1% and were classified as "responsive to connectivity" while the other 20 species were determined "non-responsive" (see Table 6). The average connectivity weight of the 25 responsive species was 27% suggesting that the explanatory power of connectivity across all species was lower than the cumulative power of the environmental variables (Figure 9). However, connectivity weight showed high variability among species ranging from 2% (*Tripleurospermum maritima*) to 100% (*Arrhenatherum elatius*).

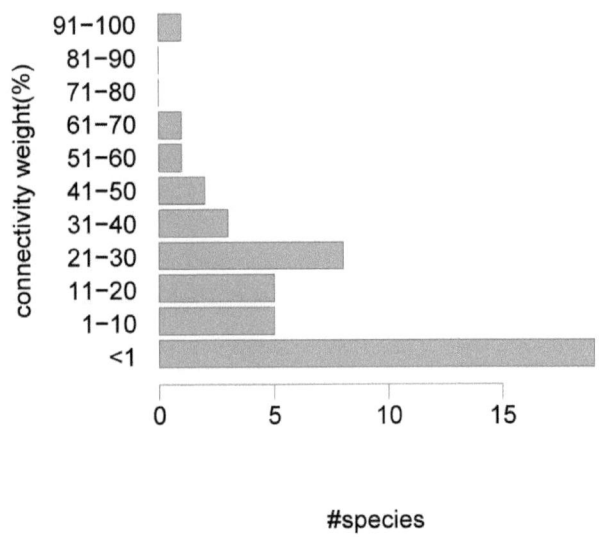

Figure 9: Frequency of connectivity weights for the studied species.

SAuC correction highly reduced the number of adequate models that entered the averaged model of a species. However, this resulted in only minor changes to the averaged models in terms of model goodness and presence of model parameters. When comparing models with and without application of the SAuC correction (steps 3 and 4 in Methods) the connectivity weights of 14 species were changed by the correction (seven decreases and seven increases, see Table 6). The connectivity weights of *Betula pendula*, *Sisymbrium altissimum* and *Taraxacum officinale* decreased below the threshold of 1.0% resulting in their reclassification from responsive to non-responsive.

4.3.2 Functional traits and connectivity

Increasing life span was positively correlated with leaf size but negatively with seed longevity index (Table 4). A negative correlation was observed between seed mass and seed number, and seed number was also positively correlated with canopy height. Terminal velocity increased with seed mass and decreased with seed number.

Table 4: Spearman correlation coefficients among plant functional traits. Significant results are in boldface

Functional trait	Canopy height	Specific leaf area	Leaf size	Life span	Terminal velocity	Seed longevity index	Seed mass
Canopy height							
Specific leaf area	-0.19						
Leaf size	0.27	-0.24					
Life span	0.03	-0.08	**0.34**				
Terminal velocity	-0.21	0.13	-0.18	-0.20			
Seed longevity index	-0.04	0.23	**-0.35**	**-0.34**	-0.02		
Log seed mass	-0.20	0.07	0.03	-0.29	**0.52**	-0.12	
Log seed number	**0.33**	-0.14	0.29	0.03	**-0.40**	0.26	**-0.56**

Species responsive to connectivity did not differ from non-responsive species in canopy height, leaf size, specific leaf area, life span or seed longevity index (Table 5). However, responsive species showed significantly lower seed numbers and higher terminal velocities than non-responsive species, while there were no differences in seed mass (Figure 10).

CHAPTER 4

Subsequent analysis revealed that the clearest distinction between connectivity response groups was obtained when combining seed number and terminal velocity in a quotient.

Table 5: Results of two-sided Wilcoxon-Mann-Whitney tests for independent samples. General difference in plant functional traits between connectivity response groups. Log (seed number/terminal velocity) is the log transformed quotient of seed number and terminal velocity. Significant results are in boldface.

Functional trait	p-value
Terminal velocity	**0.013**
Log seed number	**0.042**
Log seed mass	0.337
Canopy height	0.537
Specific leaf area	0.599
Leaf size	0.144
Seed longevity index	0.749
Log(seed number/terminal velocity)	**0.007**

4.4 Discussion

Spatial autocorrelation correction highly reduced the number of adequate models that entered the averaged model of a species. However, this resulted in only minor changes to the averaged models in terms of model goodness and presence of model parameters. When comparing models with and without application of the SAuC correction (steps 3 and 4 in Methods) the connectivity weights of 14 species were changed by the correction (seven decreases and seven increases, see Table 6). The connectivity weights of *Betula pendula*, *Sisymbrium altissimum* and *Taraxacum officinale* decreased below the threshold of 1.0% resulting in their reclassification from responsive to non-responsive.

Additionally, we found that the clearest differentiation between connectivity response groups was obtained by the combination of seed number and terminal velocity. This observation suggests a compensatory relationship between the two traits in which high dispersal range may counterbalance low numbers of dispersers and vice versa. Our results provided empirical evidence for current plant dispersal concepts, which emphasize both the role of seed number and the duration of seed transport - indicated here by terminal velocity - for dispersal across intermediate distances (Nathan *et al.* 2008).

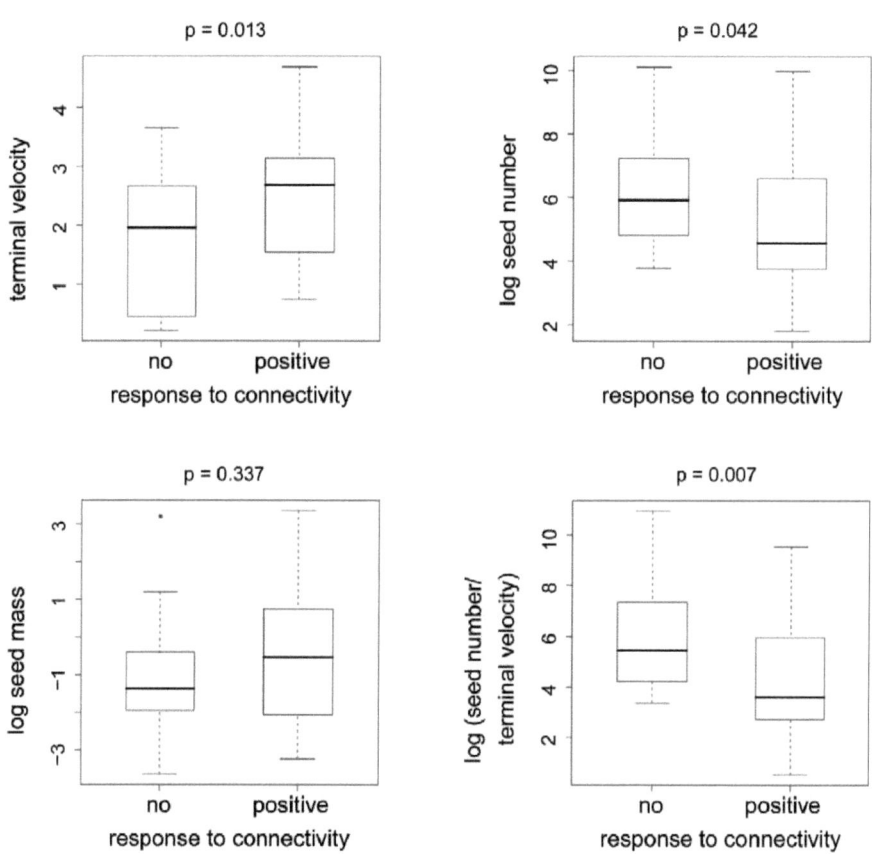

Figure 10: Relations between selected plant functional traits (terminal velocity, log seed number, log seed mass, log (seed number/terminal velocity)), and species groups based on their response to connectivity. Species were considered "responsive to connectivity" if the connectivity weight in the averaged model was >1.0. Boxplots show median, 25% and 75% quartiles, minimum and maximum, and outliers, respectively.

Seed number was also relevant in empirical studies, in which distances between habitat patches by far exceeded regular dispersal ranges (e.g. Dupré & Ehrlén 2002; Verheyen *et al.* 2004; Kolb & Diekmann 2005). But in terms of dispersal range studies have often highlighted the role of dispersal mode. Here we provide strong empirical support of wind dispersal models predicting that terminal velocity allows a quantitative assessment of dispersal capacity across intermediate distances (Tackenberg et al. 2003; Soons et al. 2004). Across larger distances, however, these mechanistic models demonstrate an increased relevance of stochastic factors (e.g. extreme winds, weather conditions). The second and third elements of our hypothesis were dismissed, because functional traits associated with local persistence were not related to plant response to connectivity. Life span, seed bank longevity, specific leaf area, canopy height or leaf size did not differ between responsive and non-responsive species. These results are inconsistent with previous empirical studies that detected relationships between fragmentation and persistence traits. For example, fragmentation vulnerability in forest ecosystems was associated with seed number, seed mass, plant height and life span (Dupré & Ehrlén 2002; Kolb & Diekmann 2005). These observations indicated the importance of functional traits associated with dispersal as well as local viability via competitive ability. By contrast, Piessens et al. (2005) showed that the ability to form persistent seed banks increased local persistence in heathland ecosystems. On national scales, Ozinga et al. (2009) found that persistent seed banks lowered the local extinction risk of plants under increased habitat fragmentation.

The particular habitat conditions and spatiotemporal dynamics of urban systems may be responsible for this discrepancy as they minimize the chance of plant recruitment from the seed bank and for local persistence via competition. New habitats for spontaneous vegetation are often created by the transformation of former agricultural land or by the abandonment of built-up spaces with impervious surfaces (Kattwinkel et al. 2009) resulting in raw soils relatively free of buried diaspores. Moreover, habitat turnover in terms of the conversion of built-up to fallow sites and vice versa occurs irregularly and with long time intervals. In other systems, such as arable fields with old soils and short disturbance return intervals due to crop rotation, or old systems with little disturbance including forests or heathlands, persistent seed banks or competitive ability may be superior to dispersal (Schippers et al. 2001). Our results suggest, however, that dispersal ability is favorable if habitat turnover is relatively slow and new habitats are created on raw soils.

Studies observing plant responses to fragmentation in old systems with a constantly increasing degree of landscape fragmentation have provided additional evidence of the confounding role of extinction debt (Grashof-Bokdam & Geertsema 1998; Helm *et al.* 2006; Ozinga *et al.* 2009). Extinction debt refers to the phenomenon that species with long life span or seed bank longevity are buffered from local extinction with the consequence that these taxa respond to habitat fragmentation with a temporal lag (Tilman *et al.* 1994). If so, distribution patterns will reflect historical rather than actual habitat configurations which may lead to an inaccurate conclusion that species with long term persistence ability are not affected by connectivity (Ewers & Didham 2006; Helm *et al.* 2006). Urban systems, however, do not demonstrate long-term trends in habitat fragmentation but are characterized by oscillating habitat connectivities due to high rates of patch destruction and creation, which may reduce the importance of extinction debt.

From a methodological point of view, our study differed from previous studies in three major aspects. First, we chose a comparable fine scale corresponding to the dispersal range of anemochorous plants. This approach increased the likelihood of discerning uncolonized habitat patches and underestimating the role of dispersal limitation.

Second, the 52 species used in this study represent a considerable fraction (>20%) and the most common species of the total species pool. This extensive species list came at the cost of considerable heterogeneity in species requirements making it difficult to develop a sampling design and select environmental variables which are meaningful to all species. Consequently, the explanatory power of species distribution models was relatively low, a characteristic commonly observed in many multi-species studies (e.g. Krauss *et al.* 2004; Kolb & Diekmann 2005).

Third, our study assessed habitat suitability for the entire study site and for each species separately. In fact, the models showed marked differences among species regarding the relative importance of connectivity and environmental predictors. These differences support the observations of previous studies suggesting that, although the average explanatory ability of connectivity may be small, single species were strongly affected (Dirnböck & Dullinger 2004; Hérault & Honnay 2005; Prugh *et al.* 2008). Furthermore, calculation of species-specific models was a necessary precondition to disentangle dispersal limitation from spatially autocorrelated habitat unsuitability. For example, *B. pendula* and *T. officinale* responded to connectivity without SAuC-correction, but showed no response after the models were corrected. *B. pendula* and *T. officinale* are well-known for their highly effective dispersal (e.g. Grime *et al.* 1989). Therefore, we assume our approach was successful in separating the effects of habitat unsuitability and dispersal limitation.

Conclusion

This study demonstrated functional relationships between species-specific connectivity and easily measured traits such as seed terminal velocity and seed number. Yet, discrepancies in trait syndromes between different published studies indicate the absence of an easy, single classification of sensitivity to fragmentation across all environments, which is hardly surprising as there is enormous variation across species and habitat types.

Furthermore, the interpretation of the results of studies conducted in different environments is also hampered by the incomparability of different methods. In particular, we want to stress the importance of appropriate spatial scales, species-specific assessment of habitat suitability, and separation of spatial autocorrelation of environmental conditions from dispersal effects. A consensus regarding methodological requirements to adequately describe the response of plants to habitat fragmentation may decidedly promote the identification of trait syndromes with increased vulnerability to fragmentation.

Trait data are becoming increasingly available for a large number of species (e.g. Kleyer *et al.* 2008) and linkages between traits and habitat configuration, as identified in our study, may allow the assessment of fragmentation vulnerability for large species data sets based on their functional traits. In the long run, this approach might be useful for the planning and design of habitat networks.

Table 1: Summary of species model averaging results and functional traits. For Nagelkerke's R^2 (R^2_N) and c presented, respectively, as obtained with and without correction for the effects of spatial autocorrelation (SAuC). I differences is given in the Methods section.

Table 6: Extended.

Species	Relative occurrence	R^2_N (with correction) SAuC-	weight (w SAuC-	weight (w SAuC-	Connecti response	Canopy h (mm)	Specific l area (mm²	Leaf size (mm²)	Life span (1: annual 2: biennia 3: perenni	Terminal velocity (Seed long index	Seed mas (mg)	Seed num
Achillea millefolium	0.50	0.24 0.24	20.7	20.7	1	323	13.35	353	3	1.41 0.11	0.11	730
Agrostis capillaris	0.25	0.16 0.16	58.8	58.8	1	249	23.03	117	3	1.21 0.66	0.04	43
Arenaria serpyllifolia	0.40	0.18 0.18	0.0	0.0	0	79	25.63	5	2	1.83 0.71	0.14	354
Arrhenater-um elatius	0.26	0.25 0.33	100.0	33.5	1	614	22.74	423	3	2.70 0.17	1.59	100
Artemisia vulgaris	0.61	0.27 0.27	20.5	20.5	1	695	17.86	1,005	3	1.55 0.76	0.13	21,474
Betula pendula	0.24	0.15 0.18	0.6	5.6	0	760	19.10	1,019	3	0.43 0.89	0.15	24,300
Bromus tectorum	0.31	0.33 0.33	0.0	0.0	0	158	25.00	136	1	2.52 0.58	2.56	74
Calamagrostis epigejos	0.72	0.24 0.25	27.0	19.7	1	619	15.01	1,093	3	0.74 0.43	0.08	288
Carex arenaria	0.28	0.26 0.26	50.3	50.3	1	95	17.39	245	3	2.85 1.00	0.59	32
Carex hirta	0.61	0.22 0.23	9.6	11.0	1	189	18.91	369	3	3.47 0.11	2.11	6
Chenopodium album	0.19	0.15 0.15	47.0	47.0	1	225	17.31	166	1	3.02 0.91	0.37	303
Cirsium arvense	0.67	0.11 0.14	0.0	0.0	0	680	8.70	706	3	0.28 0.34	1.82	69
Cirsium vulgare	0.49	0.15 0.17	0.0	0.0	0	619	12.36	1,414	2	0.58 0.29	24.49	44
Conyza canadensis	0.71	0.26 0.27	0.0	0.0	0	383	21.95	70	1	0.22 0.85	0.03	1,466
Coryne-phorus canescens	0.42	0.33 0.33	17.3	17.3	1	69	10.67	10	3	1.28 1.00	0.06	70
Dactylis glomerata	0.63	0.18 0.18	0.0	0.0	0	510	19.84	2,098	3	2.62 0.20	0.38	345
Daucus carota	0.28	0.17 0.17	34.7	34.7	1	489	16.13	978	2	2.60 0.68	1.70	929
Deschampsia cespitosa	0.52	0.17 0.17	0.0	0.0	0	363	13.07	267	3	1.48 0.27	0.09	757
Elymus repens	0.66	0.11 0.11	25.6	20.8	1	585	18.29	548	3	3.01 0.15	1.30	15
Festuca ovina agg.	0.24	0.16 0.17	35.9	35.6	1	250	16.31	116	3	2.67 0.20	0.34	62
Festuca rubra agg.	0.73	0.20 0.20	0.0	0.0	0	267	19.53	132	3	2.76 1.00	0.35	117
Hypericum perforatum	0.48	0.22 0.22	33.3	33.3	1	544	21.59	66	3	1.79 0.90	0.15	1,211
Hypochoeris radicata	0.56	0.24 0.33	0.0	0.0	0	33	16.70	583	3	0.36 0.32	3.28	178
Lotus corniculatus	0.38	0.22 0.23	21.3	18.9	1	129	15.40	117	3	3.61 0.26	6.53	21
Medicago lupulina	0.42	0.22 0.23	4.8	16.1	1	121	23.60	74	2	3.14 0.64	4.67	58
Melilotus officinalis	0.33	0.17 0.17	13.6	15.3	1	753	15.83	114	2	3.79 0.57	28.51	280
Oenothera biennis	0.66	0.11 0.11	29.4	29.3	1	627	14.64	884	2	2.69 0.88	0.62	5,865
Phalaris arundinacea	0.23	0.20 0.20	0.0	0.0	0	730	13.97	1,075	3	2.72 0.20	0.10	591

Table 6: Extended.

Species	Relative occurrence	R^2_N (with S correction). SAuC-	weight (with SAuC-	weight (with SAuC-	Connectivity response er	Canopy height (mm)	Specific leaf area (mm²/t)	Leaf size (mm²)	Life span (1: annual 2: biennial 3: perennial)	Terminal velocity (m) Seed longevity index	Seed mass (mg)	Seed number
Plantago lanceolata	0.59	0.14	0.14	14.4	25.2	1	75	18.73	708	3	3.44 0.35	1.93	52
Plantago major	0.24	0.13	0.13	0.1	0.1	0	59	17.30	2,372	3	2.73 0.79	0.80	3,005
Poa pratensis	0.65	0.16	0.17	0.0	0.0	0	315	27.31	156	3	2.09 0.39	0.20	67
Poa trivialis	0.25	0.17	0.17	66.2	66.1	1	339	28.14	119	3	1.32 0.75	0.19	151
Polygonum persicaria	0.25	0.14	0.15	5.3	45.2	1	161	17.94	315	1	3.46 0.86	5.17	31
Rumex acetosella	0.62	0.21	0.21	0.0	0.0	0	130	15.18	86	3	2.92 0.69	0.32	632
Senecio inaequidens	0.90	0.11	0.11	0.0	0.0	0	337	18.12	141	3	0.47 0.50	0.12	1,834
Sisymbrium altissimum	0.15	0.15	0.17	0.2	1.1	0	322	17.66	162	1	2.31 1.00	0.14	5,047
Tanacetum vulgare	0.71	0.26	0.26	18.6	18.6	1	514	14.48	1,796	3	2.33 0.11	0.10	1,997
Taraxacum officinale	0.56	0.23	0.25	0.0	42.6	0	69	22.21	1,553	3	0.61 0.30	0.34	391
Trifolium arvense	0.62	0.28	0.29	21.1	10.5	1	211	18.53	75	1	1.39 0.62	2.78	49
Trifolium pratense	0.34	0.32	0.32	0.0	0.0	0	294	24.16	629	3	2.43 0.32	0.19	147
Trifolium repens	0.22	0.16	0.16	0.0	0.0	0	57	26.56	263	3	3.66 0.40	0.55	128
Tripleurospermum maritima	0.42	0.29	0.29	1.8	2.7	1	458	23.12	269	2	2.33 1.00	0.13	14,700
Tussilago farfara	0.48	0.31	0.31	0.0	0.0	0	91	14.83	3,793	3	0.21 0.06	0.15	1,340
Vicia hirsuta	0.33	0.15	0.20	25.9	12.9	1	151	26.46	133	1	4.69 0.45	4.31	31
Vulpia myuros	0.49	0.24	0.25	23.2	20.7	1	118	10.36	6	1	3.05 0.25	0.31	96

Chapter 5

"I took in February three table-spoonfuls of mud from three different points, beneath water, on the edge of a little pond; this mud when dry weighed only 6 ¾ ounces; I kept it covered up in my study for six months, pulling up and counting each plant as it grew; the plants were of many kinds, and were altogether 537 in number; and yet the viscid mud was all contained in a breakfast cup."

Charles Darwin

5 Effects of the resident community on colonizing plants: A functional approach

Andrea Schleicher

Michael Kleyer

Summary

Theoretical predictions emphasizing the role of functional resemblance between colonizing and resident species are challenged by experimental evidence indicating that plant establishment is largely determined by vacant space and the colonizer's competitive response. Here we asked if different principles apply to colonizers with high and low competitive response. We expected predominantly vacant space effects for colonizers with high competitive response (weak competitors) but greater relevance of functional resemblance for colonizers with low response (strong competitors). Thereby, we tested the alternative hypotheses that colonizers are most successful if they are (a) functionally similar or (b) dissimilar to resident species. We studied establishment in a successional sere adopting a functional group approach to assess competitive response. Three different ways were considered to define functional groups from traits important for establishment and competition. For each group, establishment success was estimated from species occurrences in the seed bank and the vegetation. The relevance of vacant space and functional community composition were compared by regressions techniques. Vacant space promoted the establishment of weak and strong competitors. For both, additional effects of resident groups were observed but were less important than vacant space (average variable weight: 33%). Biotic interactions were only observed between dissimilar groups and pointed to the gradual replacement of weak by strong competitors. The establishment of weak competitors was hampered if strong competitors were abundant, whereas strong competitors were facilitated by high cover of weak competitors. Thereby, high competitive response was

Chapter 5

associated with short life span and high specific leaf area, and contrasting expressions with low response. Seed mass and canopy height regulated competitive response only across species with different life spans, but not when considering short- and long-lived species separately. We conclude that, compared to the availability of vacant space, functional resemblance played a minor role for community invasibility, irrespective of the colonizer's competitive response.

EFFECTS OF THE RESIDENT COMMUNITY ON COLONIZING PLANTS

5.1 Introduction

What is the role of biotic interactions in plant community assembly? Understanding how biotic interactions constrain the assembly of local communities from a given species pool, has been a major issue of community ecology ever since the fundamental work of Darwin (1859). Later, Diamond (1975) hypothesized that there are so-called assembly rules, i.e. general principles that describe the role of biotic interactions, and in particular competition, for observed species presences and abundances (Wilson & Gitay 1995).

In a plant functional framework, the functional resemblance between colonizing and resident species may influence colonization success. Colonizers may be most successful if they are either functionally similar or dissimilar to the residents. According to the limiting similarity theory, only those plants should be able to establish that are functionally dissimilar to the extant community (Elton 1958; MacArthur & Levins 1967). Residents may reject entering species through the preemption of available resources (Naeem et al. 2000; Dukes 2001) if the traits of residents and colonizers enable similar resource use (Fargione et al. 2003; Mwangi et al. 2007). By contrast, functional similarity may promote colonization under the assumption that successful colonizers and established species have to be similar in those traits that govern plant response to environmental conditions (Levine 2000).

These predictions are however challenged by experimental evidence describing colonization as a competitive weighted lottery (Mouquet et al. 2004). Establishment is then largely determined by the availability of bare ground with reduced competition and the colonizer's response to competition (e.g. Burke & Grime 1996; Davis et al. 2000; Thompson et al. 2001). Plants with high competitive response may enter resident communities with high competitive effect only if disturbance has cleared space of competitors (Ehrlén & van Groenendael 1998; Schippers et al. 2001). Trait expressions such as small seeds, reduced canopy height and annual life span are considered as indicators of high competitive response (Grime 2002). By contrast, tall perennials producing large seeds exhibit low competitive response and are capable to establish in closed vegetation (Leishman 1999; Liancourt et al. 2009). Competitive response is further associated with species ability to deplete light and soil resources which is a matter of canopy height and growth rate (Weiher et al. 1999; Díaz et al. 2004; Violle et al. 2009). Thereby, growth rate is closely associated with specific leaf area (SLA) (Garnier 1992; Reich et al. 1992). Additionally, the ability to generate storage organs and differences in life span have been suggested important to further differentiate general resource capture strategies under varying levels of resource availability (i.e. stress) (Eriksson 1996; Klimeš *et al.* 1997).

In this study, we investigated the role of functional resemblance between colonizing and resident species differing in competitive response. We expected that species with traits indicating high competitive response (weak competitors) are not able to enter communities with closed canopies but largely depend on bare ground. By contrast, functional resemblance with resident species should be important for colonizers exhibiting traits associated with low competitive response (strong competitors). To address functional resemblance we considered the two alternative hypotheses that establishing plants are most successful if they are (i) functionally similar or (ii) functionally dissimilar to the resident community.

Assembly rules based on functional resemblance are often deduced from co-occurrence patterns of species across communities. The analysis of patterns is, however, no direct evidence for the generating mechanism because community assembly may have taken place under other than the present environmental conditions (Chase *et al.* 2005; Holyoak *et al.* 2005). Assembly rules can be assessed more reliably when the colonization of plants in extant communities is investigated. Thereby, colonization success depends on the colonizer's ability to arrive in a locality, and subsequently, to germinate and to establish. Here, we focused on the establishment success of successfully arrived colonizers by comparing the species compositions of seed banks and extant communities. For simplicity, we refer to colonizers and colonization success to describe the species found in the seed banks and their establishment probabilities, although we omitted the case of dispersal limitation.

As a model system, we explored successional sequences starting from unvegetated deposits of raw sandy sediments in an urban environment. Successions are particularly suited for such analyses because the decline of bare ground in favor of plant cover may be regarded as a gradient of competitive pressure (Peet 1992; Weiher & Keddy 1999). We used a functional group approach to assess the relative importance of vacant space versus resemblance with the resident community for colonizer success of groups with high and low competitive response. Three possibilities to define functional groups were considered based

EFFECTS OF THE RESIDENT COMMUNITY ON COLONIZING PLANTS

on traits important for plant capacities to establish, to exploit soil resources and to compete for light. We then calculated for each group of potential colonizers a measure of colonization success based on species presences in the local seed bank and in the vegetation of the following season. We compared the explanatory powers of bare ground availability and resident functional groups similar and dissimilar to the colonizer group to assess the role of functional resemblance between colonizers and residents for establishment success.

5.2 Methods

5.2.1 Study site

The study was conducted at an industrial park in the city of Bremen, Germany (53°05′N, 8°44′E, mean annual temperature 8.8°C, mean annual precipitation 694 mm: Deutscher Wetterdienst 2006/07) representing an artificial island habitat. The area was created by filling the original marshland with about 2 m of sand in a step-by-step procedure beginning in the 1970s. By 2007, this process had generated a patch network of about 4.0 km^2 and an age gradient spanning nearly 40 years. The majority of the area was used for development, including building construction and infrastructure. However, a substantial fraction was never utilized and ruderal communities established. The eight most abundant species in 2007 were (in decreasing order) *Senecio inaequidens*, *Holcus lanatus*, *Rumex acetosella*, *Arenaria serpyllifolia*, *Poa pratensis*, *Vulpia myuros*, *P. trivialis*, and *Agrostis capillaris* (nomenclature following Jäger & Werner 2002).

We applied a stratified random sampling method with strata including the entire patch age range from zero to nearly forty years as an estimate of successional progress. Patch age was determined by an aerial photos time series available at 5 to 10-year intervals since 1972. Patch age was processed in a GIS (ESRI Inc. 2006), and patches were subsequently assigned to eight age classes resolved by the analysis. Finally, we placed equal numbers of plots per age class at random coordinates.

5.2.2 Establishment success and failure

To assess establishment success and failure we compared species presence in the seed bank with the emergence of new species at a given plot in the subsequent vegetation period. To assess seed bank composition we collected sediment cores in early March 2008 assuming that natural stratification had taken place over the winter. At each plot we randomly took six

CHAPTER 5

replicates of 10 cm deep, cylindrical soil cores of 4 cm diameter which were subdivided into three horizons of 0-2 cm, 2-5 cm and 5-10 cm depth (Bakker 1989). The cores of each plot were pooled per horizon and stored in plastic bags at 1°C in the dark until germination. Germination trial largely followed Ter Heerdt *et al.* (1996). In May, we spread each sample to an even depth in a plastic tray (60 x 40 x 6 cm) previously lined with water-soaked, sterile fleece. As soil grain size approximated seed size of expected species we decided against sieving and removed roots and tillers by hand. The trays were arranged randomly in the greenhouse and watered daily. Seedlings were counted weekly and then removed to prevent overcrowding. Seedlings that could not be identified in a tray were transplanted to empty pots and allowed to grow until they could be identified. After 16 weeks, no new seedlings emerged and the trial was terminated. Most seeds were found in the uppermost horizon and, given that recruitment from deep horizons was unlikely (Milberg 1995; Fenner & Thompson 2004), we constrained subsequent analyses on the seed bank composition of the uppermost 2 cm layer.

To determine the emergence of new species in a given plot we recorded the plants in each plot in summer 2007 and in summer 2008, i.e. after seed bank sampling which took place in early spring 2008. We subdivided each plot (1 m^2) into 100 subplots of 0.1 x 0.1 m and counted the presence of each vascular plant species in the subplots.

Establishment success was defined as absence in the established vegetation in 2007, but presence in the seed bank and established vegetation of 2008. Failure was defined as absence in the established vegetation in 2007 and 2008, and presence in the seed bank of 2008.

In the study site, bare soil resulted from the recent land fillings which were not yet completely covered with vegetation. That is, the proportion of bare soil indicated successional progress. Additionally, small-scale above-ground damages to the vegetation arose from digging activities of rodents, or from temporary material deposition and traffic during the industrial development of the area. We estimated the percentage of bare soil, recorded signs of disturbances that occurred between the summers 2007 and 2008 and used them as two separate explanatory variables.

5.2.3 Assessment of abiotic habitat unsuitability

Establishment failure is not necessarily contingent on inhibitory effects of the resident community. It may as well be explained by abiotic habitat conditions that are unsuitable for a potential colonizer.

To assess species habitat preferences we investigated environmental conditions at each plot by collecting soil samples at each soil horizon to a depth of 0.8 m. No plant roots were noted below that depth. The following soil parameters were determined in the laboratory: plant available potassium (flame photometer, Egnér *et al.* 1960), phosphorus (Continuous Flow Analyser, Murphy & Riley 1962), pH (gauged in a $CaCl_2$- solution), calcium carbonate ($CaCO_3$) and bulk density (all according to Schlichting *et al.* 1995). Soil texture, soil moisture and organic content were measured in the field according to the standards in Ad-hoc-AG Boden (2005). Soil aeration was derived from bulk density, and soil texture according to empirical functions in Ad-hoc-AG Boden (2005).

We subsequently performed an outlying mean index analysis (OMI: Dolédec *et al.* 2000) on the environmental variables to determine habitat suitability for each species in the regional pool. OMI, unlike constrained correspondence analysis, makes no assumptions on the length of gradients and assigns equal weight to species-poor and species-rich sites. Therefore, it is well-suited to investigate multidimensional niches. As ordination techniques are often more sensitive to skewed data than to non-normal data distributions (Legendre & Legendre 2006) we applied power transformations to soil aeration, pH, and plant available potassium and phosphorous to yield skewness between -1 and 1 (McCune & Grace 2002). Nearly 50% of the plots were free of calcium carbonate. Therefore, calcium carbonate was recoded as a categorical variable (with/without calcium carbonate).

5.2.4 Functional group definitions

Competitive effect and competitive response (i.e. the depletion of resources due to plant activity and plant response to a change in local resources due to competition: Violle *et al.* 2009) are often related to the same traits but associated with contrasting expressions (Violle *et al.* 2009). For instance, large canopy height is associated with low competitive response and high competitive effect. Therefore, we will not distinguish between response and effect traits in the following.

Functional groups were defined at the base of the functional variation in the total species pool, i.e. vascular plants that have been observed in the study site between 2003 and 2008 (255 species in total). We considered three possibilities to define functional groups (i.e. three groupings), each separating the species pool into three to four functional groups that differed in competitive response traits (see Figure 11). A first grouping ("establishment grouping") was based on traits relevant to establishment (seed mass, canopy height, SLA, life span). The "resource capture grouping" considered traits relevant to resource capture and turnover (canopy height, SLA, presence of storage organs, life span). Finally, the "strategy grouping" comprised information on seed mass, canopy height and SLA because those have been suggested to differentiate leading plant ecological strategies including competitive response (Westoby et al. 2002).

For each grouping, species of the regional species pool were categorized into functional groups by means of multivariate clustering. The correlative structures among traits can result in the overestimation of the effective dimensionality of the functional space in which species separate (Petchey & Gaston 2002). Therefore, we first decreased dimensionality by calculating a principal component analysis from of the standardized traits using the R-command "dudi.mix" from the ade4 package which can deal with mixed quantitative variables and factors (Thioulouse et al. 1997). Seed mass and canopy height were log-transformed to reduce the effects of species with extreme values. Then we applied Ward's Minimum Variance clustering with the Calinsky-Harabasz stopping criterion on the resulting species scores to construct functional groups.

For each functional group we calculated the cumulative cover from the frequencies of resident species belonging to that group in that locality in 2008. Newly established species were omitted from this calculation to avoid circularity in subsequent regression analyses. Functional groups can reach frequencies >100, thus overemphasizing species-rich functional groups, when simply summarizing species frequencies. Therefore, we calculated cumulative cover following Tichý & Holt (2002). Canopy height, SLA, life span, and seed mass values for all species were extracted from the LEDA Traitbase (Kleyer et al. 2008), whereas presence of storage organs was taken from Clo-Pla 3 (Klimešová & Klimeš 2006).

5.2.5 Statistical analyses

For a given functional group, we first calculated establishment success and failure as the sum of seed bank abundances of species belonging to that functional group that respectively succeeded and failed to establish from the seed bank. From the obtained values, we calculated the proportion of establishment success.

We examined general differences in establishment success proportions among functional groups of a certain grouping with Kruskal-Wallis-tests as ANOVA-assumptions were not fulfilled. Finally, we used logistic regression analysis to predict establishment success of colonizing functional groups from the following predictor variables: (i) disturbance, (ii) bare soil and (iii) the covers of functional groups present in the resident community. For each functional group, only groups of the same grouping were considered, i.e. the relevance of functional groups of the establishment grouping was only tested for colonizers of the establishment grouping, and so on. Establishment proportions relying on a single occurrence in the seed bank were downweighted in the logistic regression to reduce the likelihood of spurious findings. Bare soil and, when necessary, functional group covers were log-transformed to minimize unequal variances.

We applied model averaging as proposed by Burnham & Anderson (2002) in preference to more conventional methods such as stepwise selection procedures. Contrary to such procedures, model averaging does not rely on the arbitrary choice of a threshold (typically 0.05 or 0.1) to judge the significance of an explanatory variable. Instead, the relative importance of each explanatory variable is derived from its consistent presence in significant models (Sanderson *et al.* 2005). Model averaging has increasingly been used in ecological studies (e.g. Sanderson *et al.* 2005; Strauss & Biedermann 2006; Kattwinkel *et al.* 2009).

CHAPTER 5

In detail, we first determined the shape of the relationship with all explanatory variables. We constructed univariate models in which relationships could be either sigmoid or unimodal, and used a likelihood-ratio test to decide upon which model to keep for subsequent analysis (Strauss & Biedermann 2006). Then, we calculated univariate and multiple models for all possible combinations of variables for a given functional grouping. This yielded 31 or 63 candidate models for each functional group depending on whether the corresponding grouping comprised three or four groups. From these candidate models, we picked a set of adequate models that met the following criteria (Strauss & Biedermann 2006): First, all coefficients differed significantly from zero ($p < 0.05$). Second, the model performed better than any model of a lower hierarchy according to a likelihood-ratio test ($p < 0.05$). If residual overdispersion was detected in a model we refitted the model with a quasibinomial family to assess the significance of the relationship (Crawley 2005).

Figure 5: Overview of trait expressions of the functional groups (capital letters below figures) of three different group definitions (groupings: "Establishment", "Resource capture", "Strategy"). Presented are boxplots (continuous traits) and stacked barplots (ordered and binary traits). Boxplots show median, 25% and 75% quartiles, minimum and maximum, and outliers, respectively. Different small letters (above figures) indicate that the functional groups differed significantly in their trait expressions according to Kruskal-Wallis rank sum tests (continuous traits) and Chi-squared contingency table tests (ordered and binary traits). Parentheses indicate that the functional trait was not considered for functional group definition.

EFFECTS OF THE RESIDENT COMMUNITY ON COLONIZING PLANTS

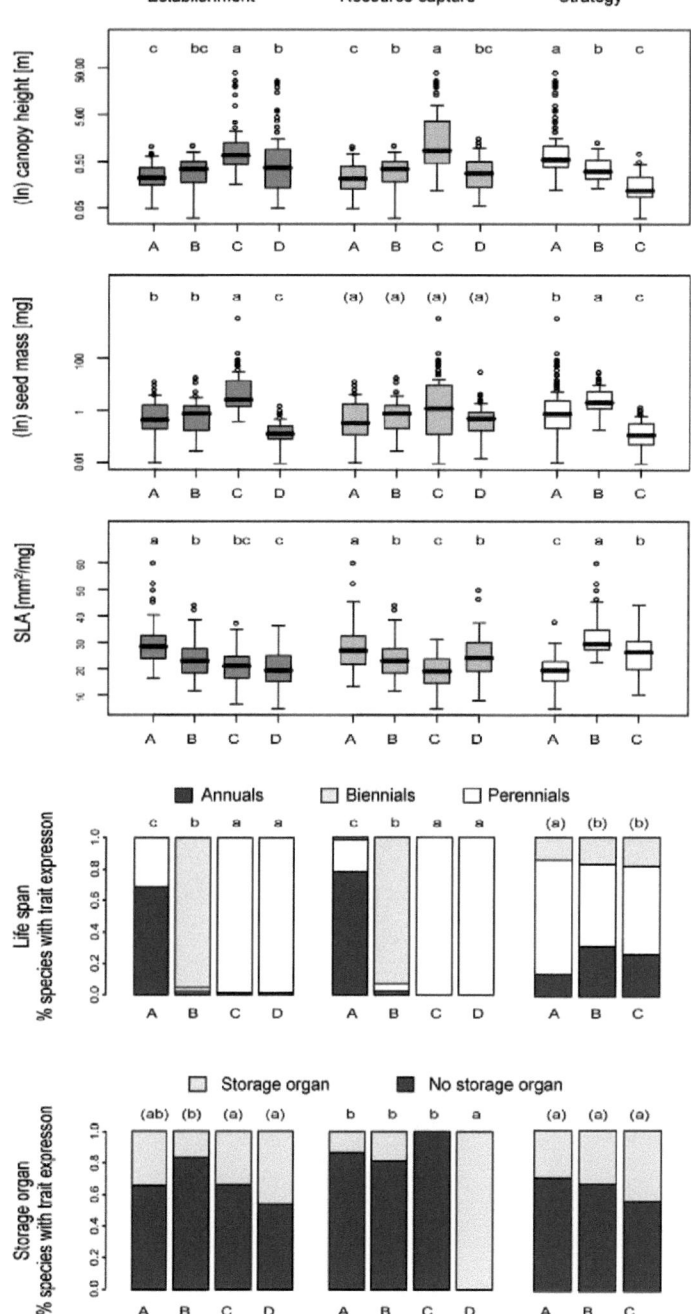

CHAPTER 5

Subsequently, we applied model averaging (Burnham & Anderson 2002) to merge adequate models into one averaged model per functional group. Next to an averaged coefficient, model averaging yields for each variable a variable weight which can be used to determine the relative importance of explanatory variables (Burnham & Anderson 2002). Accordingly, we used these variable weights to compare the importances of vacant space on the one hand and functional composition of the resident community on the other hand. All statistical analyses were performed in R (R Development Core Team 2005).

Table 7: Summary of soil variables measured at the plot level, with their units, means, standard deviations where applicable, and their eigenvector scores at the 1st axis of the outlying mean index analysis (OMI). Organic content was classified in four levels (0: 0; 1: <1; 2: 1 - <2; 3: 2 - <4; 4: 4 - <8 mass%), soil moisture in five levels (0: > 4.0; 1: 4.0 - >2.7 2: 2.7 - >2.1; 3: 2.1 - >1.4; 4: <=1.4; 5: 0 lg hPa), and $CaCO_3$ was treated as categorical variable with two levels (with/without $CaCO_3$).

	Unit	Minimum	Median	Maximum	Correlation coefficient with 1st OMI axis
Soil aeration	[volume %]	15.19	32.00	38.00	0.58
Phosphorous content	[kg/ha]	0.00	181.10	2,876.00	-0.34
Potassium content	[kg/ha]	0.00	189.90	895.30	-0.33
pH		4.48	6.39	7.70	-0.23
Organic content	Ordinal [mass-%]	0	1	4	-0.48
Soil moisture	Ordinal [lg hPa]	0	2	5	-0.04
$CaCO_3$ content	Categorical	Not applicable	Not applicable	Not applicable	0.41

* Levels following Ad-hoc-AG Boden (2005)

5.3 Results

5.3.1 Results of the outlying mean analysis

The first two axes of the OMI analysis accounted for 42% and 22% of total variation in marginality, i.e. the variation in the distances between the mean soil conditions used by species and the mean soil conditions of all plots. The first axis was primarily related to soil aeration and the second to soil pH (Table 7). The Monte-Carlo randomization test was significant at $p < 0.001$ for the mean marginality. However, only 14 out of 134 species showed a significant deviation of their niche from the average soil conditions suggesting that soil parameters influenced only a minority of species (Figure 12). Consequently, we assumed that abiotic habitat unsuitability was of minor importance for governing establishment failure.

5.3.2 Plant functional groups

An overview over trait expressions distinguishing the functional groups of the three groupings is given in Figure 11. The establishment and the resource capture grouping clearly differentiated annual and biennial life span groups (A and B). The two groupings differed, however, in the distinction of perennials. Perennials in groups C and D of the establishment grouping contrasted in seed mass, whereas the groups C and D of the resource capture grouping differed in generation of storage organs. The strategy grouping discriminated three functional groups. Group A consisted of tall species with low SLA, group B species with high values of SLA and seed mass, and species in group C combined small canopy height and low seed mass. The groups A of the resource capture and the establishment groupings, and the group C of the strategy grouping were characterized by comparably short life span, low canopy height, small seeds and high SLA.

5.3.3 Taxonomic and functional diversity

Altogether 124 species were identified during vegetation surveys in 2008 with local species richness in plots averaging 16 ± 6 species. Functional group richness varied little among plots with most functional groups being represented in each locality by at least one species. Only group C of the establishment grouping (perennials with high canopy height and seed mass) was missing in more than 20% of the plots. Mean functional group covers ranged

CHAPTER 5

between 5% and 21% (groups B and A of the strategy grouping), i.e. a single group never covered more than a quarter of the area of a plot.

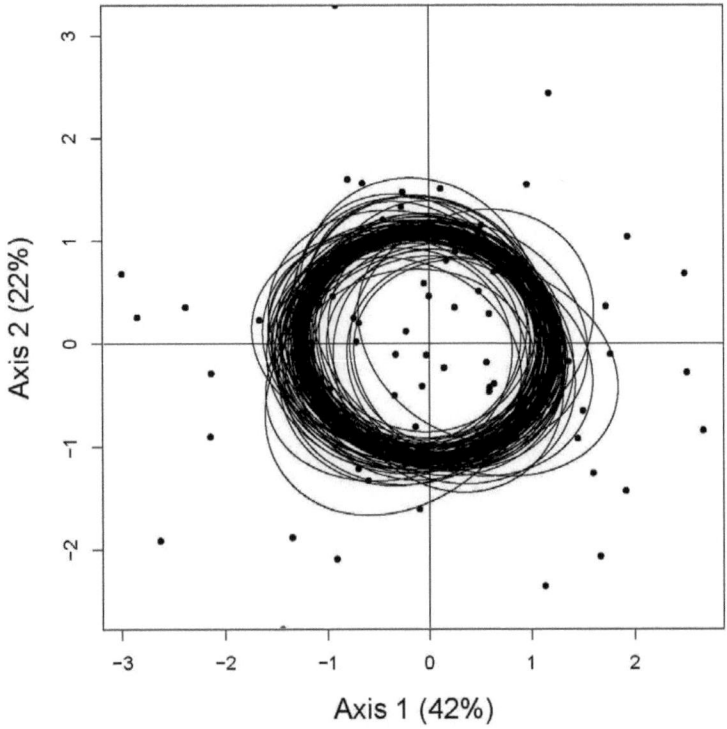

Figure 12: Outlying mean index analysis (OMI) of 134 plant species observed in the study area on local environmental variables. Axes inertias are given in parentheses, for interpretation of the axes see Methods. Dots represent plots and ellipses are a graphical summary of the cloud of points of each species representing the environmental niche of each species. The center of each ellipse is centered on the means, its width and height are given by the variances, and the covariance sets the slope of the main axis of the ellipse.

EFFECTS OF THE RESIDENT COMMUNITY ON COLONIZING PLANTS

The species found in the seed bank represented the most prevalent species in the established vegetation. 59 species were present in the seed bank with an average of 11 ± 5 species per plot. Species occurred on average in only 19% of the sampled seed banks demonstrating large differences in seed bank composition among plots. Consequently, not all functional groups were represented in the seed bank of each plot, resulting in different sample sizes of functional groups (range: 27-76 out of 77). In the resulting data sets, however, all functional groups were composed of at least two species per plot.

There were no differences in establishment proportions among functional groups of the resource capture grouping (Table 8). However, we observed smaller establishment proportions for group C of the establishment, and group B of the strategy grouping compared to other groups of the same grouping. Yet, these two groups differed in their competitive responses as suggested by differences in canopy height, SLA, life span und storage organs (Figure 11).

Table 8: Results of the Kruskal-Wallis rank sum tests for general differences in establishment success rates among functional groups of three different group definitions. If appropriate, post-hoc pairwise comparisons were executed and letters indicate which groups differ significantly ("a" being attributed to the group(s) with the highest median value and "c" to the group with the lowest median value).

Grouping	df	n	X^2		Functional group			
					A	B	C	D
"Establishment"	3	42	13.25	**	a	ab	b	a
"Resource capture"	3	46	3.53		a	a	a	a
"Strategy"	2	25	7.81	*	a	b	a	

* $p < 0.05$
** $p < 0.01$

5.3.4 Relevance of vacant space and resident community for establishment success

Recent damages to the vegetation canopy were noted in 21 out of 77 plots. Proportion of bare soil ranged between 0 and 99.5% (median: 9.6%). Disturbed plots showed higher percentages of bare soil than undisturbed (p =0.007, one-sided Wilcoxon-Test). Correlations between bare soil and the covers of functional groups never exceeded Spearman's rho =0.6.

Significant relationships with disturbance or bare ground were detected for all functional groups except group B of the strategy grouping (large seeds and high SLA; Table 9). For this group no significant model could be achieved. Establishment success of most other functional groups was higher after recent disturbances of the resident vegetation. Only three functional groups did not respond to disturbance, or were little affected by disturbance (variable weight <7%). These groups were characterized by short life span and low canopy height. Two groups largely consisted of annuals with high SLA (groups A of the establishment and the resource capture groupings), whereas group C of the strategy grouping comprised species with intermediate SLA producing small seeds. These three groups established most successfully when the proportion of bare ground was high as shown by variable weights of bare soil >80% (see Table 9).

The functional group composition of the resident community affected the establishment success of only few colonizer groups. Negative and positive relationships can be distinguished. Group C of the strategy grouping established less successfully at increased cover of the most dissimilar species (i.e. tall long-lived species with large seeds; group A). Given a variable weight of 11%, however, the explanatory value of this relationship was limited. Positive relationships were detected for three functional groups. Groups D, C, and A of the establishment, resource capture, and strategy grouping, respectively, were taller and exhibited larger seeds, longer life span and lower SLA than the facilitating resident functional groups (groups B, B, and C of the same groupings). These resident groups had in common that they showed the lowest SLA and highest life span values of their grouping. Canopy height, seed mass and presence of storage organs showed no consistent pattern among these groups.

Table 9: Summary of the relationships between the establishment success of functional groups and the availability of vacant space (disturbance and % bare soil), and the covers of functional groups in the resident community. Three different ways to distinguish functional groups were investigated, denoted as "Strategy", "Resource capture" and "Establishment" grouping. Each grouping comprised three or four functional groups indicated as Group A, B, C, or D. Presented are median trait values of the colonizing functional groups for canopy height (CH, in m), specific leaf area (SLA, in mm^2/mg) and seed mass (SM, in mg), as well as Nagelkerke's R^2 (R^2_N), and, for each explanatory variable in the regressions, parameter estimates and variable weights (in brackets) as obtained from model averaging. "n.s." denotes no significant relationships.

Colonizing group	Median trait values			R^2_N	Disturbance	% Bare ground	Re...	
	CH	SLA	SM				Group A	Group
"Strategy"								
Group A	0.55	19.5	0.72	0.27	1.30 [74]	n.s.[0]	n.s.[0]	n.s.[0]
Group B	0.31	29.5	2.00	-	n.s.[0]	n.s.[0]	n.s.[0]	n.s.[0]
Group C	0.12	26.4	0.11	0.17	0.05 [6]	0.32 [83]	-0.06 [11]	n.s.[0]
"Resource capture"								
Group A	0.22	27.0	0.32	0.31	n.s.[0]	0.58 [100]	n.s.[0]	n.s.[0]
Group B	0.35	23.2	0.75	0.21	1.59 [100]	n.s.[0]	n.s.[0]	n.s.[0]
Group C	0.86	19.2	1.17	0.48	2.04 [51]	n.s.[0]	n.s.[0]	0.74 [49]
Group D	0.28	24.2	0.48	0.30	1.84 [100]	n.s.[0]	n.s.[0]	n.s.[0]
"Establishment"								
Group A	0.22	28.5	0.42	0.32	n.s.[0]	0.58 [100]	n.s.[0]	n.s.[0]
Group B	0.34	23.2	0.73	0.26	1.73 [100]	n.s.[0]	n.s.[0]	n.s.[0]
Group C	0.67	21.2	2.50	0.30	1.22 [38]	-0.67 [62]	n.s.[0]	n.s.[0]
Group D	0.37	19.5	0.12	0.42	1.96 [55]	0.00 [0]	n.s.[0]	0.38 [45]

CHAPTER 5

Resident functional groups only affected colonizing groups when they were dissimilar. Colonizer groups were never facilitated by residents from the same group.

5.4 Discussion

Predicting community assembly requires understanding the relative importance of vacant space and resident species for colonizer success. Here, we asked whether different principles apply to colonizers with traits associated with high and low competitive response. Our initial expectation was that the availability of space determined the colonization success of weak competitors, whereas functional resemblance with resident species should be more relevant for strong competitors. Regarding functional resemblance, we tested the two alternative hypotheses that (i) functional similarity, and (ii) functional dissimilarity, enhanced establishment success. However, our results indicated that, irrespective of the colonizer's competitive response, bare ground was the main driver of establishment success. Yet, resident species' effects were also found, and they influenced weak and strong competitors differently. Short life span and high SLA were indicative for high competitive response. Hence, life span and SLA seemed to control competitive response. By contrast, the relevance of seed mass and canopy height depended on the colonizer's life span. The ability to generate storage organs was not important to differentiate species with high and low competitive response.

5.4.1 Establishment of colonizers with high competitive response

Our first hypothesis was that the establishment of weak competitors was determined by the availabilty of bare ground. Indeed, functional groups composed of annuals responded exclusively to the proportion of bare ground. This has been observed for annuals before (Bullock 2000; Grime 2002) but has often been attributed to their small seeds (e.g. Burke & Grime 1996; Kalamees & Zobel 2002; Fenner & Thompson 2004). In fact, seeds in short-lived species are not always smaller than in perennials (Shipley & Peters 1990). In our study, groups dominated by annuals did not exhibit smaller seeds than other groups. Annuals showed however larger SLA than perennials. High SLA is associated with maximal potential growth rates of seedlings in full light which allows annuals to grow and reproduce rapidly (Marañón & Grubb 1993; Westoby et al. 1996). High growth rates are however often associated with high respiration rates (Grime 1965) which may induce strongly negative growth rates when light availability falls below the compensation point

(Mahmoud & Grime 1974). Consequently, annuals with high SLA suffer the greatest reduction of seedling growth rates in shade (Fenner & Thompson 2004) which may exclude them from later successional stages with closed canopies (Fenner 1978).

Yet, weak competitors were not always unaffected by resident species. In the strategy grouping, where groups were not well differentiated by life span, small colonizers with small seeds were negatively affected by taller and larger-seeded residents. Small-seeded species are particularly disadvantaged by dense vegetation because their germination requirements conflict with established species effects on microclimatic conditions (reduced light intensities, altered light spectra, diminished daily fluctuations in temperature and moisture: Fenner & Thompson 2004). Similarly, small canopy height has been associated with increased seedling sensitivity against shadow casting species, and particularly late in the growing season when perennial species form closed canopies (Anten & Hirose 1999).

Our results suggested that variation in seed mass and canopy height had little control over establishment success of plants with similar life span. Only when annuals, biennials and perennials were pooled small seed mass and canopy height appeared to determine high competitive response. This result is in line with other studies stating that life cycle differences are major predictors of establishment success and deserve more attention in community ecological studies (Shipley & Parent 1991; Roscher *et al.* 2009).

5.4.2 Establishment of colonizers with low competitive response

For colonizers with traits indicating low competitive response, we had hypothesized an enhanced relevance of resident species, and in particular, of functional resemblance with resident species. This was only partly confirmed.

Strong competitors were stronger influenced by resident species than weak competitors (variable weights 45 and 39% versus 11%, see Table 9). They established better if weaker competitors were abundant in the resident community, namely shorter-lived species with higher SLA. This relationship fits typical successional sequences, during that short-lived species able to exploit resources rapidly are replaced by slower growing perennials (Clements 1916; Connell & Slatyer 1977). Our observation further adds to the increasing evidence that not competition alone, but also facilitation, is important during community assembly (e.g. Prieur-Richard *et al.* 2000; von Holle & Simberloff 2004; Roscher *et al.* 2009).

However, disturbance of the resident vegetation was also important for the colonization success of strong competitors. In fact, bare ground and disturbance were always more important for invader success than resident functional groups, irrespective of the competitive response of the colonizers. While the variable weights of bare ground and disturbance exceeded 50%, the weights of resident functional groups averaged only 29%. The overriding importance of the availability of bare soil has often been acknowledged (e.g. Burke & Grime 1996; Symstad 2000). Yet, to our knowledge only Roscher *et al.* (2009) investigated the relative importance of bare soil and the functional composition of the resident community. Consistent with our observation, they reported that the establishment success of most species groups was mainly explained by bare soil, rather than the prevalence of functional groups (36.5% versus at most 7%). However, Roscher *et al.* (2009) did not differ among colonizers with high and low competitive response.

In our study, weak competitors were promoted by bare soil, but strong competitors by disturbance. As disturbance we denoted any above-ground, comparatively small-sized damages to the vegetation sward. Therefore, disturbance may be interpreted as the (temporary) reduction of competitive pressure associated with an increased availability of light and soil resources. Contrarily, the proportion of bare soil was related to successional age. Maximum bare soil occurred in the youngest plots. Accordingly, disturbed and open plots may have differed in microclimatic conditions as well as light availability at the soil surface. Diminished levels of light in disturbed plots may have promoted the establishment of strong competitors with increased growth rates in the shade (low SLA), whereas weak competitors with high SLA thrived in young plots with ample bare ground.

5.4.3 Role of functional resemblance

Is establishment promoted if colonizers are functionally similar or if they are dissimilar to the resident community? The hypothesis of functional similarity reducing the invasibility of a community is rooted in the idea that resident species can lower the availability of local resources (light, nutrients, water) to a level which inhibits the establishment of further individuals (MacArthur & Levins 1967; Fargione *et al.* 2003). We did not detect any relationship, negative or positive, between establishing and established species of the same functional group. Instead, all significant relationships referred to dissimilar groups. Positive interactions between dissimilar groups might be taken as indirect evidence for limiting similarity. But this was opposed by our results showing that dissimilarity only operated one-directional, i.e. weak competitors were affected by strong competitors but not vice versa. That is, from the perspective of colonizers with high competitive response functional dissimilarity hampered rather than promoted establishment.

Our results join the row of studies that found no evidence for limiting similarity governing the assembly of plant communities (e.g. Symstad 2000; von Holle & Simberloff 2004; Losure et al. 2007; Mwangi et al. 2007; Roscher et al. 2009). The most comprehensive support for limiting similarity effects was provided by a study of Fargione et al. (2003) who observed that negative interactions were always strongest between species of the same functional group. Yet, they also reported that the strongest effects for all colonizer groups were exerted by a single functional group with the greatest capacity to lower the limiting resource in that system. This fits our observation of superior competitors replacing inferior species, as well as other studies reporting that presence of a single dominant species group may be more important for community invasibility than functional resemblance with the entering species (Prieur-Richard et al. 2000; Symstad 2000; Mwangi et al. 2007).

The results of studies using a functional group approach are strongly dependent on the selection of traits (Díaz & Cabido 1997; Petchey & Gaston 2002). To tackle this problem we considered several functional groupings composed of different traits (Lavorel et al. 1999; Thompson et al. 2009). We concluded that traits showing the most consistent statistical evidence across all different groupings were most important. These traits were life span and SLA. In contrast, high seed mass and canopy height were characteristic for strong competitors in the strategy grouping but not in the resource capture grouping, which seemed to limit their general relevance.

As a conclusion, resident species influenced the establishment of weak and strong competitors. Thereby, biotic interactions were only observed when colonizing and resident species were functionally dissimilar. Yet, the type of biotic interaction differed between colonizers with traits indicating high and low competitive response. The establishment of weak competitors was hampered by the presence of stronger competitors, whereas strong competitors established better when weak competitors were abundant. In the long run this probably resulted in the replacement of weak by stronger competitors. Moreover, we found that disturbance may facilitate mainly the entry and establishment of strong competitors. This has consequences for the management of metacommunities composed of weak and strong competitors. If disturbances facilitate the colonization of communities composed of weak competitors by strong competitors, conservation measures such as mowing may demonstrate inappropriate to preserve such communities. All in all, bare ground and disturbance were most important predictors of colonization success, suggesting that functional resemblance between colonizers and residents had limited relevance for community assembly.

Chapter 6

"If several (such) investigations yield no significant results, it will be difficult to avoid the conclusion that plants are not too interested in ecological theory."

Anni J. Watkins & J. Bastow Wilson

6 Functional patterns during succession: Is plant community assembly trait-driven?

Andrea Schleicher

Michael Kleyer

Cord Peppler-Lisbach

Summary

Little is known about changes in the significance of functional filtering and neutral processes during the succession of plant communities. Generally, a shift is predicted from dispersal processes in initial phases to competition filtering in later stages. In this study, we assumed if community assembly is trait-driven, the shift is reflected in functional community composition, whereas random functional patterns indicate trait-neutral mechanisms. Therefore, we tested for filter-induced convergence and divergence in urban plant communities along a successional gradient. We investigated dispersion of traits associated with species capacities to disperse, tolerate abiotic stress, and compete at two scales (100 x 100 cm and 10 x 10 cm) and compared the results to null model expectations. We subsequently used regression trees to associate convergence and divergence to plot age and stress due to low water and soil nutrient availability. Significant patterns were not detected for most traits, and the variance explained by the regression trees was often lower than 20%. Instead of the expected convergence towards high seed number, low seed terminal velocity and annual life span during early succession, we found divergence in seed number. Stress filtering played a minor role which was explained by an outlying mean analysis on soil parameters indicating that most plants occupied similar environmental niches. Convergence in seed bank longevity occurred at intermediate plot age, and convergence in the combination of life span and lateral spread demonstrated the relevance of competition filtering in the most fertile plots. However, competition-induced convergence in canopy

CHAPTER 6

height or specific leaf area was not supported by our results. Overall, we concluded that the evidence was against strong filtering of traits but for trait-neutral community assembly during the first 40 years of succession.

6.1 Introduction

The question of how plant communities are assembled has resulted in nearly a century of unresolved debate on the predictive value of assembly rules (Clements 1916; Weiher & Keddy 1999; Hubbell 2001). On the one hand, community assembly is considered as a filtering process the outcome of which is predictable from species functional traits. In this deterministic model, species are selected from a regionally available pool if species capacities for dispersal, stress and competitive tolerance meet the local requirements as exposed by spatiotemporal isolation, environmental conditions and community invasibility (Keddy 1992; Lavorel & Garnier 2002). On the other hand, concepts emphasizing trait-neutral assembly have challenged this view as these concepts emphasize the importance of arrival sequence, priority effects or demographic stochasticity (see Robinson & Dickerson 1987; Hubbell 2001). In such a framework, community composition is explained as the result of highly random processes – rather than a consequence of predictable functional filtering.

Currently, the ecological consensus is that filter and neutral mechanisms simultaneously affect community assembly (Hubbell 2001; Leibold & McPeek 2006). However, the relative importance of either mechanism is hypothesized to vary among ecosystems and even different stages of community formation (Leibold *et al.* 2004; Cottenie 2005). During the course of succession, a shift in filters' relevances that determine local plant diversity is expected. While neutral or trait-driven dispersal processes prevail primarily in the initial phase of succession, competition filtering elicits more of a response as succession advances and dense canopies develop (Peet 1992; Weiher & Keddy 1999).

Direct tests of neutrality require detailed species and (meta-) community information which is often difficult to obtain (Nee & Stone 2003; Gotelli & McGill 2006). Alternatively, trait-neutral processes may be indirectly indicated by random trait distributions, whereas non-random distributions designate filtering processes that shape functional community composition each in a characteristic way (Grime 2006; Mason *et al.* 2007). Consequently, comparing observed patterns with random expectations may allow the evaluation of trait-driven versus trait-neutral assembly mechanisms (Gotelli & McGill 2006).

CHAPTER 6

How do dispersal, stress and competition filters shape community assembly during succession? Functional convergence and divergence may be considered as the most likely non-random patterns (e.g. MacArthur & Levins 1967; Losos 2000; Grime 2006). Convergence and divergence denote that the functional similarity of species forming a community will be respectively greater and smaller than the corresponding similarity of a random assembly. This can be explained by strong selective pressure placed on a trait by each filter, which sieves only small subsets of the total range of trait expressions available from the geographical species pool (Keddy 1992; Díaz et al. 1999). In contrast, random trait distributions are expected if trait-neutral processes prevail.

Dispersal filtering is assumed high if a locality is exposed to strong spatial or temporal isolation generating convergence towards increased dispersal ability. Divergence may be induced if alternative dispersal strategies are equally successful and manifest themselves in contrasting trait expressions. However, a single dispersal strategy is often favored. For example, wind dispersal has repeatedly been shown as the main dispersal vector early in succession (Fenner 1987; Prach & Pyšek 1999), whereas bird dispersal plays a major role in shrub communities (Miles 1987; Ozinga et al. 2004). If only the dispersal filter is operative, traits associated with stress tolerance and competitive ability should not be convergent or divergent, but randomly distributed.

Similar to dispersal filtering, environmental stress may induce convergence. Yet, stress filtering is not restricted to early successional stages, and involves stress tolerance traits rather than dispersal traits. The existence of a tree line, i.e. convergence towards non-phanerophytic life forms (Tranquillini 1979; Wielgolaski & Karlsen 2007) is only one prominent example. Alternative adaptive strategies to manage environmental stress are reported for many ecosystems (Larcher 1984; Grime 2002), increasing the expectation to detect divergence. In dry grasslands, for example, the co-occurrence of ephemerals and scleromorphic perennials may result in the generation of divergence in traits associated with growth and reproductive rates.

Predictions of functional patterns induced by the competition filter are probably the most controversial. Functional divergence in competition related traits is predicted by theories of niche differentiation and character displacement (Brown & Wilson 1956; Hutchinson 1959). Based on the assumption that competition is most intense among species with identical resource exploitation and associated trait expressions, coexistence should require a minimum dissimilarity in resource use which exposes an upper limit to the functional similarity of interacting species (i.e. limiting similarity: MacArthur & Levins 1967; Pacala & Tilman 1994). In contrast, functional convergence is considered as the result of competition if more superior competitors exclude inferior species from a community (Tilman 1990; Grime 2002).

The competition filter, different from the dispersal and the stress filter, acts only among neighboring species and therefore requires a consideration of scale (Huston 1999; Murrell & Law 2003; Stubbs & Wilson 2004). Competitive effects may only be detected over very small distances, in the case of herbaceous plants, typically not exceeding a few centimeters (Wilson & Whittaker 1995; Holdaway & Sparrow 2006). Only late in succession, when competitive exclusion has already resulted in a community-wide absence of inferiors, may the competition filter elevate to the community level.

The necessity to consider functional patterns at the community and neighborhood scale is often regarded as adding unpleasant complexity to the investigation of assembly rules (Lawton 1999; Weiher & Keddy 1999). However, it provides an opportunity to separate competition filtering from dispersal and stress filtering. We assume that functional convergence or divergence displayed at the community scale indicates stress filtering, whereas competition filtering should be observable at the neighborhood scale, unless convergence toward traits associated with high competitive effects represents exclusion of inferior species.

An alternative approach to disentangle stress and community filters resides in the application of constrained null models (Peres-Neto *et al.* 2001). Trait convergence and/or divergence in a community are usually investigated by null models, which compare observed estimates of functional diversity with estimates of random assemblies. Constrained null models generate random communities by constraining randomization according to species aptitudes to occur in habitats with specific abiotic conditions, which differs from ordinary null models. Clearly, this approach provides best results in the presence of strong environmental gradients.

CHAPTER 6

In this study, we explored the importance of trait-driven relative to trait-neutral community assembly by the investigation of functional patterns, i.e. convergence and divergence in a successional series. We judged the importance of trait-driven filter processes by the presence of functional patterns in corresponding traits and scales. The absence of non-random functional patterns was considered as evidence for trait-neutral mechanisms (see Table 10). We expected the following results:

- Convergence in dispersal traits during the initial phase of succession at the community scale.
- Convergence or divergence in stress tolerance traits during the entire successional gradient. Divergence indicates alternative plant strategies to effectively manage environmental stress.
- Convergence or divergence in later successional stages in traits related to competitive ability at the neighborhood scale. Convergence is predicted to result from the presence of a superior competitor group, whereas divergence points to limiting similarity. At later stages of succession, these patterns eventually elevate to the community scale.
- Random functional patterns indicate trait-neutral assembly.

Therefore, we determined the presence of functional convergence and divergence at the community and neighborhood scales, using a null model approach by comparing observed functional patterns with random expectations. In addition to analyzing contrasting functional patterns at two scales, we calculated constrained null models to discern stress from competition filtering. The functional patterns identified in the model were subsequently related to successional age and stress level estimated to determine the importance of assembly filters in different parts of the successional and the environmental gradients. This allowed us to track variation in the importance of trait-driven relative to trait-neutral mechanisms during the course of succession.

Table 10: Overview of hypothesized functional patterns generated by trait-driven and trait-neutral assembly mechanisms at the community and neighborhood scales. Capitals denote convergence (C), divergence (D) and random trait distributions (R).

Scale	Trait-driven assembly			Trait-neutral assembly
	Dispersal filter	Stress filter	Competition filter	
Community	C in dispersal traits	C in stress tolerance traits: a single strategy to cope with stress D: alternative strategies	R C in competitive effect traits: competitive exclusion has already taken place	R
Neighborhood	R	R	C in competitive effect traits: one superior functional group D: limiting similarity	R

6.2 Methods

6.2.1 Study area

The study was conducted at an industrial park in the city of Bremen, Germany (53°05′N, 8°44′E, mean annual temperature 8.8°C, mean annual precipitation 694 mm: Deutscher Wetterdienst 2006/07) representing an artificial island habitat. The area was created by filling the original marshland with approximately 2 m of sand in a step-by-step procedure beginning in the 1970s. By 2007, this process had generated a patch network of about 4.0 km² and an age gradient spanning nearly 40 years. Consequently, the area differed from the surrounding landscape in edaphic conditions and elevation, which induced sharp contrasts in water and soil resource availability, and naturally, species pools.

The majority of the area was used for development, including building construction and infrastructure. However, a substantial fraction was never utilized and ruderal communities established. Seed bank analyses indicated that the foreign fill material was almost free of diaspores (data not shown). Therefore, we are confident that dispersal rather than regeneration from the soil seed bank was the dominant colonization mechanism.

We applied a stratified random sampling method with strata including the entire patch age range from zero to nearly forty years as an estimate of successional progress. Patch age was determined by an aerial photos time series available at 5 to 10-year intervals since 1972. Patch age was processed in a GIS (ESRI Inc. 2006), and patches were subsequently

CHAPTER 6

assigned to eight age classes resolved by the analysis. Finally, we placed equal numbers of plots per age class at random coordinates.

6.2.2 Vegetation data

Ruderal plant communities dominated the study area, with the following eight most abundant species recorded in 2007 (in decreasing order): *Senecio inaequidens, Holcus lanatus, Rumex acetosella, Arenaria serpyllifolia, Poa pratensis, Vulpia myuros, P. trivialis*, and *Agrostis capillaris* (nomenclature following Jäger & Werner 2002).

Vegetation composition data was collected at the local community and neighborhood scales. We sampled 92 plots and counted the frequencies of each vascular plant species in 100 subplots of 0.1 x 0.1 m totaling 1 m². We subsequently used species frequencies from the 1 m² plot for the local community scale analysis, and species composition per subplot for the neighborhood scale analysis. In addition, we were interested in plant coexistence and evaluated species presence/absence according to the vertical projection of its above-ground parts on the ground surface rather than its rooting point.

6.2.3 Trait selection

Estimates of functional divergence are highly sensitive to the identities and ranges of traits (Cadotte *et al.* 2009). Nevertheless, many studies have explored functional divergence in a single a priori selected trait or in functional volume spanned by all available traits associated with the function of interest (e.g. Kraft *et al.* 2008; Cornwell & Ackerly 2009; Thompson *et al.* 2009). In this study, we followed a different approach. For each filter, we selected particular traits based on current knowledge of trait-environment relationships. We considered seven single traits and five combinations of traits associated with dispersal capacity, tolerance to abiotic stress and competitive ability (see Table 13). In the following, trait combinations are denoted by an "&" between involved traits.

Dispersal ability was described as follows: the number of seeds, the seed number and terminal velocity quotient, and the seed number, specific leaf area (SLA) and life span combination. High capacity for anemochorous dispersal has been associated with high seed number (Kolb & Diekmann 2005) and its additive effect with terminal velocity (Grashof-Bokdam & Geertsema 1998; Tackenberg *et al.* 2003, also see Chapter 4). SLA and life span serve to distinguish two types of pioneer species. Moderate growth rates often characterize long-lived species. However, short-lived species require increased growth rates to rapidly generate many seeds (Grubb 1987; Vitousek & Walker 1987; Grime 2002). SLA was used as an indicator of plant growth rate (Garnier 1992; Reich *et al.* 1999).

Depending on the limiting resource and its temporal variability, stress tolerance may involve a variety of functional attributes. These include reduced SLA (Fonseca *et al.* 2000; Wright *et al.* 2004a), annual life form (Weiher *et al.* 1999; Bossuyt & Honnay 2006), production of persistent seed banks or the generation of clonal organs to store or translocate resources (Warner & Chesson 1985; Klimeš *et al.* 1997). Additionally, we considered the combination of a persistent seed bank and resource translocation via clonal organs explain a trade-off between traits that enable plants to compensate periods with reduced resource availability in time or space (Eriksson 1996; Ehrlén & van Groenendael 1998).

Competitive ability describes a species capacity to capture resources in the presence of neighbors. Depending on whether competition is for soil resources or light, increased competitive ability is attributed to high SLA values (Garnier 1992; Reich *et al.* 1992) or canopy height (Anten & Hirose 1999; Gross *et al.* 2007). Moreover, increased life span or high lateral spread have both been associated with space acquisition (Gaudet & Keddy 1988; Gross *et al.* 2007) and may therefore be considered as alternative strategies. Finally, increased competitive effect has been observed for plants that exhibit long life span and late flowering (Grime 2002). We used a combination of life span and flowering month as well as life span and lateral spread to investigate interactive effects.

Information on canopy height, SLA, life span, lateral spread, seed bank longevity and seed number were extracted from the LEDA Traitbase (Kleyer *et al.* 2008), and the first month of flowering from Jäger & Werner (2002). We used lateral spread to assess the mere existence of clonal organs (lateral spread >0) as well as the distance resources may be translocated by clonal organs.

6.2.4 Assessment of functional convergence and divergence

CHAPTER 6

The choice of index can be crucial when exploring the distribution of coexisting species in functional space (Stubbs & Wilson 2004). Recent studies suggested breaking down functional diversity into functional richness, evenness, divergence and dispersion (Villéger et al. 2008; Laliberté & Legendre 2010). These statistics allow dealing with any number of traits, different data types (i.e. quantitative, semi-quantitative and qualitative) and weighting of species and traits.

Figure 6: Flowchart of the computational steps performed for each trait and trait combination to assess significant functional convergence and divergence, at the community (plot: pale grey area) and at the neighborhood (subplot: dark grey area) scale. Significance is derived from the percentage of random dispersions greater (convergence) or smaller (divergence) than the observed dispersion (p <0.05). Random dispersions were generated by swapping the rows of species x trait matrix in a null model algorithm with 1,000 randomizations (white rectangle).

FUNCTIONAL PATTERNS DURING SUCCESSION

Our interest was in the distribution of trait values, and therefore, we evaluated functional

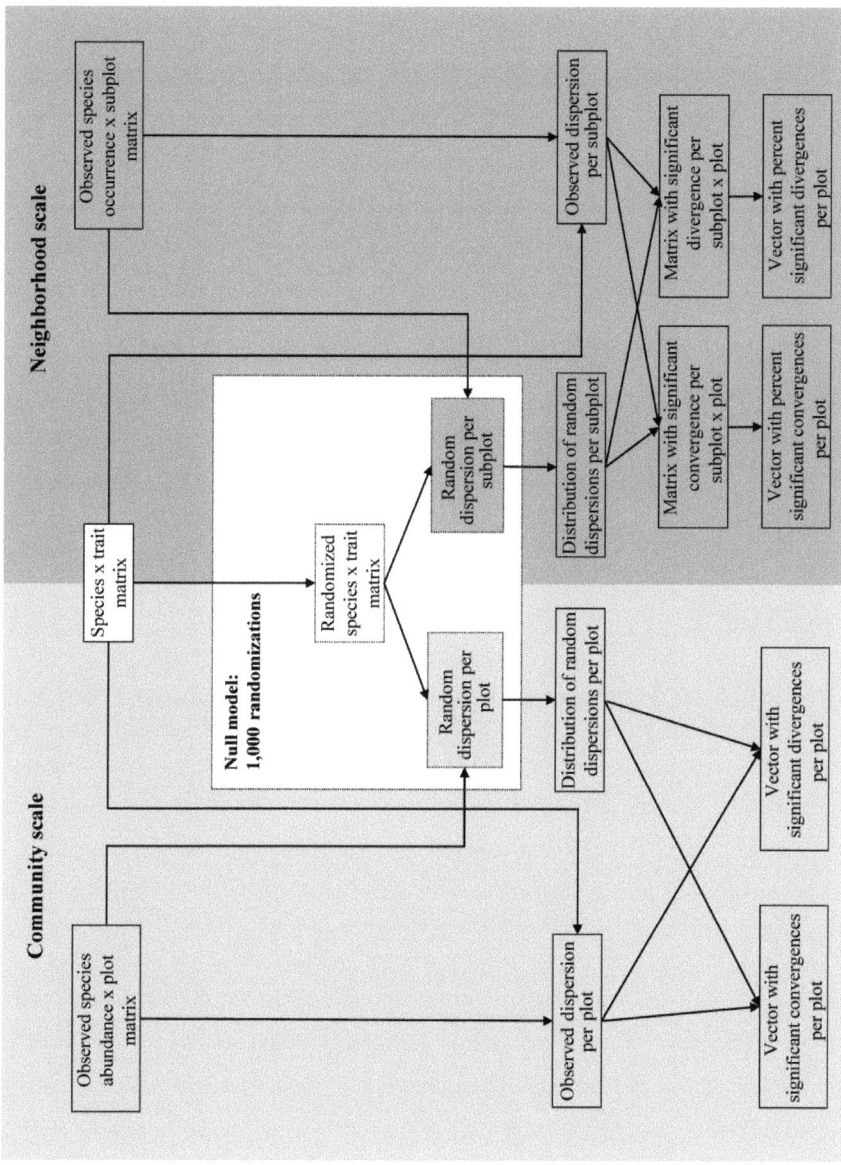

dispersion (Laliberté & Legendre 2010). Functional dispersion (FDis) describes the dispersion or spread of trait expressions within the functional volume occupied by the community. But contrary to functional divergence as defined by Villéger et al. (2008), FDis incorporates information about the size of that volume (Laliberté & Legendre 2010), which is relevant when comparing communities. FDis may therefore be regarded as a surrogate of Villéger et al.'s measures of functional richness and functional divergence, combining the advantages of both statistics. Preliminary analyses indicated that functional richness was highly redundant with FDis, whereas functional evenness and divergence had little meaning when dealing with low species numbers (n <6), which we observed at the neighborhood scale (0.1 x 0.1 m). Therefore, functional evenness and divergence data are not presented here.

To preserve realistic functional variation in the species data set we used largely untransformed trait data. The only exceptions were seed number and canopy height which were log-transformed to reduce the effects of species with extreme trait expressions. To calculate FDis of trait combinations, the traits selected for analysis were standardized to range between zero and one to assign traits equal weights.

The comparison of observed values with corresponding values from a random assembly, using a null model approach, was applied to assess the significance of functional patterns. We followed Watkins & Wilson (2003) and Stubbs & Wilson (2004) to randomize the rows of the species x trait matrix rather than species x plot matrix to generate null model communities, i.e. species were randomly assigned the functional characteristics of another species in the species pool without replacement. Therefore, by maintaining the observed trait-correlation structure, biological realism within the randomized communities was preserved.

FDis for all single traits and trait combinations was calculated for all observations, and for 1,000 randomizations of each plot (local community scale), as well as each subplot (neighborhood scale; see Figure 13). The proportion of randomizations with a test statistic equal to that observed, or more extreme, gave the probability of the observed FDis under the null model (Mason et al. 2007; Mason et al. 2008). If the observed FDis was lower than expected at random with a threshold of $p < 0.05$, this indicated that competition had induced significant functional convergence, and if the observed FDis was higher than expected at random, this indicated significant divergence. At the community scale, FDis was weighted by species frequencies in each plot.

6.2.5 Constrained null model approach

To assess habitat preferences of species in the regional pool, we investigated environmental conditions at each plot by collecting soil samples at each soil horizon to a depth of 0.8 m. No plant roots were noted below that depth. The following soil parameters were determined in the laboratory: plant available potassium (flame photometer, Egnér et al. 1960), phosphorus (Continuous Flow Analyser, Murphy & Riley 1962), pH (gauged in a $CaCl_2$-solution), calcium carbonate ($CaCO_3$), and bulk density (all according to Schlichting et al. 1995). Soil texture, moisture and organic content were measured in the field according to the Ad-hoc-AG Boden (2005) standards. Soil aeration was derived from bulk density, and soil texture according to empirical functions in Ad-hoc-AG Boden (2005).

We subsequently performed an outlying mean index analysis (OMI: Dolédec et al. 2000) on the environmental variables to determine habitat suitability for each species in the regional pool. OMI, unlike constrained correspondence analysis, makes no assumptions on the length of gradients and assign equal weight to species-poor and species-rich sites. Therefore, it is well-suited to investigate multidimensional niches.

Ordination techniques are often more sensitive to skewed data than to non-normal data distributions (Legendre & Legendre 2006). Therefore, we applied power transformations to soil aeration, pH, and available potassium and phosphorous to yield skewness between -1 and 1 (McCune & Grace 2002). Nearly 50% of the plots were free of calcium carbonate. Therefore, calcium carbonate was recoded as a categorical variable.

CHAPTER 6

6.2.6 Analyses of the relationships between functional patterns and gradients of succession and stress

Significant functional patterns (trait convergence, divergence) may be limited to parts of the successional gradient or certain levels of resource depletion. This requires a technique that can detect nonlinear relationships, as well as small sections along the environmental gradient with trait convergence or divergence. We used regression tree analysis (Breiman *et al.* 1984) to uncover the relationship, if any, between (i) presence/absence of functional patterns at the community scale, and (ii) percentage of subplots with significant functional convergence or divergence per plot to plot age and soil aeration (see Figure 13). Regression trees were built with a minimum bucket size of three and cross-validated with 100 randomizations.

Relationships among functional traits were assessed with Spearman correlations for continuous traits and Wilcoxon-Mann-Whitney tests for categorical variables. All statistical analyses were performed using the software R (R Development Core Team 2005) and the packages "ADE4" (OMI analysis, Thioulouse *et al.* 1997), and "rpart" (Regression Tree Analysis, Therneau *et al.* 2009). Null models were generated by an adapted version of the command "oecosimu" in the package "vegan" (Oksanen *et al.* 2009), and functional dispersions were calculated with the function "dbFD" (Laliberté & Legendre 2010).

6.3 Results

6.3.1 Taxonomic and functional diversity

Across all plots we identified 134 species representing the total species pool. The species exhibited high variability in functional trait expressions. For instance, seed number and canopy height varied by eight and three orders of magnitude, respectively (Table 11). This variability was also reflected in a substantial variation of FDis among plots (Figure 14).

Correlations among traits were generally weak and only correlations with Spearman coefficients >0.3 are discussed. The strongest correlations were observed between canopy height and seed number, and life span and lateral spread (r_s =0.47 and r_s =0.38, respectively). Taller species were also characterized by lower SLA (r_s = -0.31), and seed number was inversely correlated with terminal velocity (r_s = -0.34). Moreover, species with short-term the seed bank persistence exhibited higher seed numbers and later flowering months than species forming transient seed banks (Wilcoxon-Mann-Whitney test: p =0.04, and p =0.03).

Table 11: Overview of plant traits and trait combinations, and functional importance. Letters indicate that the trait (combination) is relevant for dispersal capacity (D), competitive capacity (C), or stress tolerance (S). Trait combinations are denoted by an "&" between involved traits.

Functional trait	Unit	Minimum	Median	Maximum	Functional importance
Seed number	[ ]	3	735	12,410,000	D
Terminal velocity	[m/sec]	0.07	1.87	5.00	D
Flowering month	[ ]	1	6	8	C
Canopy height	[m]	0.03	0.36	40.00	C
SLA	[mm^2/mg]	4.90	23.35	60.09	C/S
Lateral spread	ordinal				C/S
	(0: no clonal organ; 1: <1 cm; 2: 1-25 cm; 3: >25 cm)				
Life span	ordinal (1: annual; 2: biennial; 3: perennial)				C/S
Seed bank longevity	categorical (0: transient; 1: short-term persistent)				S
Seed number & SLA & life span					D
Seed number / terminal velocity					D
Flowering month & life span					C
Life span & lateral spread					C/S
Lateral spread & seed bank longevity					S

CHAPTER 6

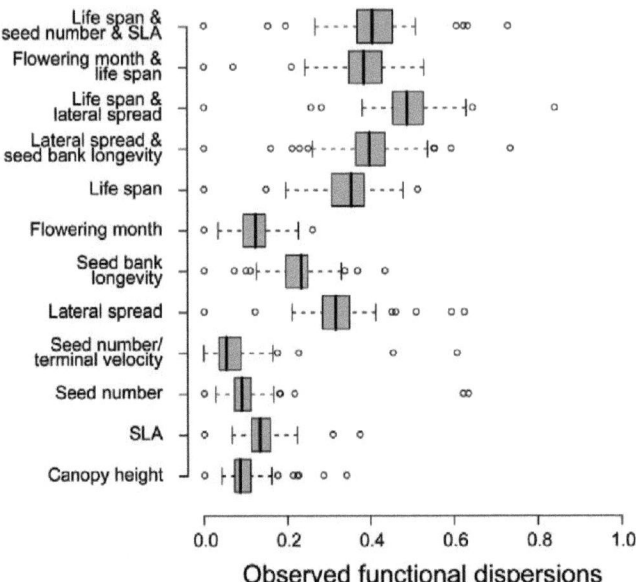

Figure 14: Boxplots of observed functional dispersions (FDis) for single functional traits and trait combinations observed at the community scale. Boxplots show median, 25% and 75% quartils, minimum and maximum, and outliers, respectively.

6.3.2 Results of the outlying mean analysis

The first two axes of the OMI analysis accounted for 42% and 22% of the total variation in marginality, i.e. the variation in distances between (i) the mean habitat conditions used by a species, and (ii) the mean habitat conditions of all plots. The first axis was primarily related to a gradient in soil resource availability as explained by soil aeration (Table 12). The second axis was primarily explained by soil pH. The Monte-Carlo randomization test was significant at $p < 0.001$ for mean marginality. However, only 14 out of 134 species showed a significant deviation of their niche from the average soil conditions, suggesting that soil parameters influenced only a minority of the species (Figure 15). Consequently, in the environmentally constrained null models only the relative occurrence probabilities of the 14 species varied among plots, and the remaining 120 species were assigned equal probabilities.

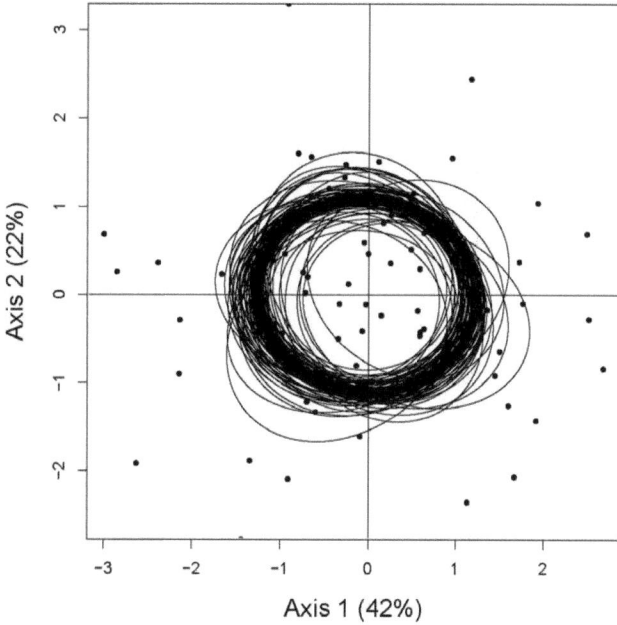

Figure 15: Outlying mean index analysis (OMI) of 134 plant species observed in the study area on local environmental variables. Axes inertias are given in parentheses. Axes interpretations are provided in Methods. Dots represent plots and ellipses are a graphical summary of the cloud of points generated by each species, which represents each species environmental niche. The center of each ellipse is anchored on the mean, its width and height are established by the variance, and the covariance sets the main axis slope of the ellipse.

The results of the OMI supported the absence of a strong environmental filter in our system. Indeed, constrained null models provided no qualitative departure from the results of the unconstrained algorithm, and we will only describe the results of the unconstrained null models in the following text. However, the OMI demonstrated that among all soil variables, soil aeration contributed most to the explained variance. Therefore, we considered soil aeration as a viable estimate for the variation in physical stress among plots during subsequent analyses.

Table 12: Summary of soil variables measured at the plot level, with units, means, standard deviations where applicable, and correlation coefficients with the 1st axis of the outlying mean index analysis (OMI). NA denotes not applicable.

Soil variable	Unit	Minimum	Median	Maximum	OMI correlation
Soil aeration	[volume %]	15.19	32.00	38.00	0.58
Phosphorous	[kg/ha]	0.00	181.10	2,876.00	-0.34
Potassium	[kg/ha]	0.00	189.90	895.30	-0.33
pH	[ ]	4.48	6.39	7.70	-0.23
Organic content [1]	Ordinal [mass-%]	0	1	4	-0.48
Soil moisture [1]	Ordinal [lg hPa]	0	2	5	-0.04
$CaCO_3$ [2]	Categorical	NA	NA	NA	0.41

[1] Levels following Ad-hoc-AG Boden (2005): Organic content: 0: 0%; 1: <1%; 2: 1- <2%; 3: 2- <4%; 4: 4- <8%. Soil moisture: 0: >4.0 lg hPa; 1: 4.0- >2.7 lg hPa; 2: 2.7-> 2.1 lg hPa; 3: 2.1- >1.4 lg hPa; 4: <=1.4 lg hPa; 5: 0 lg hPa
[2] two levels: with/without $CaCO_3$

6.3.3 Functional community patterns

Comparison of observed and null model dispersions provided more support for trait-neutral than trait-driven assembly processes. The results indicated that most observed FDis values were congruent with the values generated by the null model, both at the community and neighborhood scale (Figure 16). At the community scale, convergence in seed bank longevity and the combination lateral spread & seed bank longevity occurred in 37% and 14% of the plots, respectively, whereas the remaining traits showed significant functional convergence in at most 4% of the plots (Figure 16 A). Divergence was more common than convergence. Divergence was observed in canopy height, seed number and flowering month (in 52, 35 and 30% of the plots, respectively), while other traits diverged in at most 6% of the plots (Figure 16 C).

FUNCTIONAL PATTERNS DURING SUCCESSION

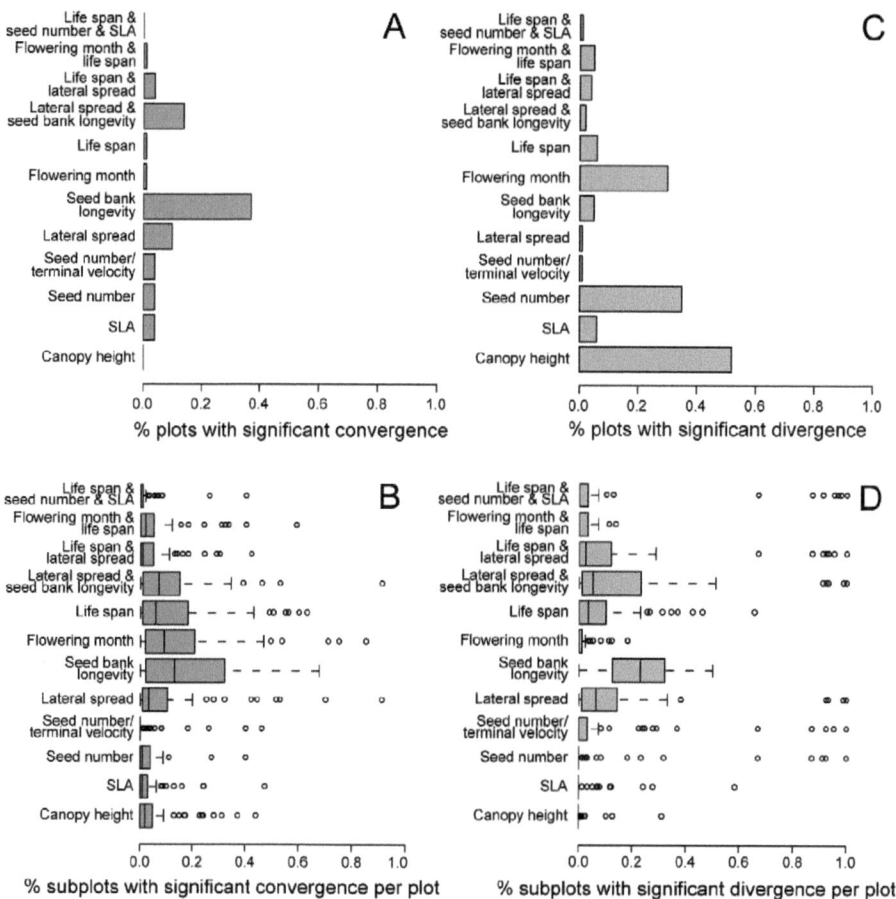

Figure 16: Frequencies of significant functional convergence (A, B) and divergence (C, D) at the community (plot) and the neighborhood (subplot) scale. Bars in figures A and B represent fractions of plots with significant convergence and divergence observed at the community scale. Boxplots in figures C and D show percentages of subplots with significant convergence and divergence per plot observed at the neighborhood scale. Indicated are median, 25% and 75% quartiles, minimum and maximum, and outliers, respectively.

At the neighborhood scale we observed high variability within plots. In seed bank longevity, for example, there was significant divergence in almost 100% of the subplots in some plots, but average percentage of the subplots with divergence was 22% (Figure 16 D). Similarly,

CHAPTER 6

the percent convergence in seed number averaged only 3%, but in some plots, 100% showed significant convergence (Figure 16 B).

6.3.4 Occurrence of functional convergence and divergence along gradients of succession and stress

The results of the regression tree analyses suggested weak associations between functional patterns and succession and stress gradients. Convergence and divergence in some traits corresponded to distinct successional stages and levels of soil aeration, however, the variation explained by the gradients was generally low (<48% at the community scale, <28% at the neighborhood scale, Table 13). In the following, we only report results where regression trees explained at least 20% of the variation (Figure 17).

At the community level, convergence towards transient seed banks was restricted to intermediate successional stages (median community weighted mean: 1.05), i.e. convergence did not occur in plots older than 26 or younger than 18 years (median community weighted mean: 1.35, compared by one-sided Wilcoxon rank test, $p = 0.01$, Figure 17 A). A strong relationship between divergence and succession and stress gradients at the community level was not detected.

Figure 7: Regression trees indicating the relationship between functional convergence and divergence to plot age (age) and soil aeration (SAer). The response variables were the occurrence of functional convergence observed at the community scale (A), and the frequencies of functional convergence (D) or divergence (B, C, E) at the neighborhood scale per plot. Respective traits or trait combinations are indicated in the headline. Each of the splits is labeled with the variable and its values (age in years, SAer in vol%) that determine the split. Each node is labeled with the percentage of observations (A) or the mean frequency (B - E) and number of observations in the group. % variation explained is indicated at the bottom of each tree.

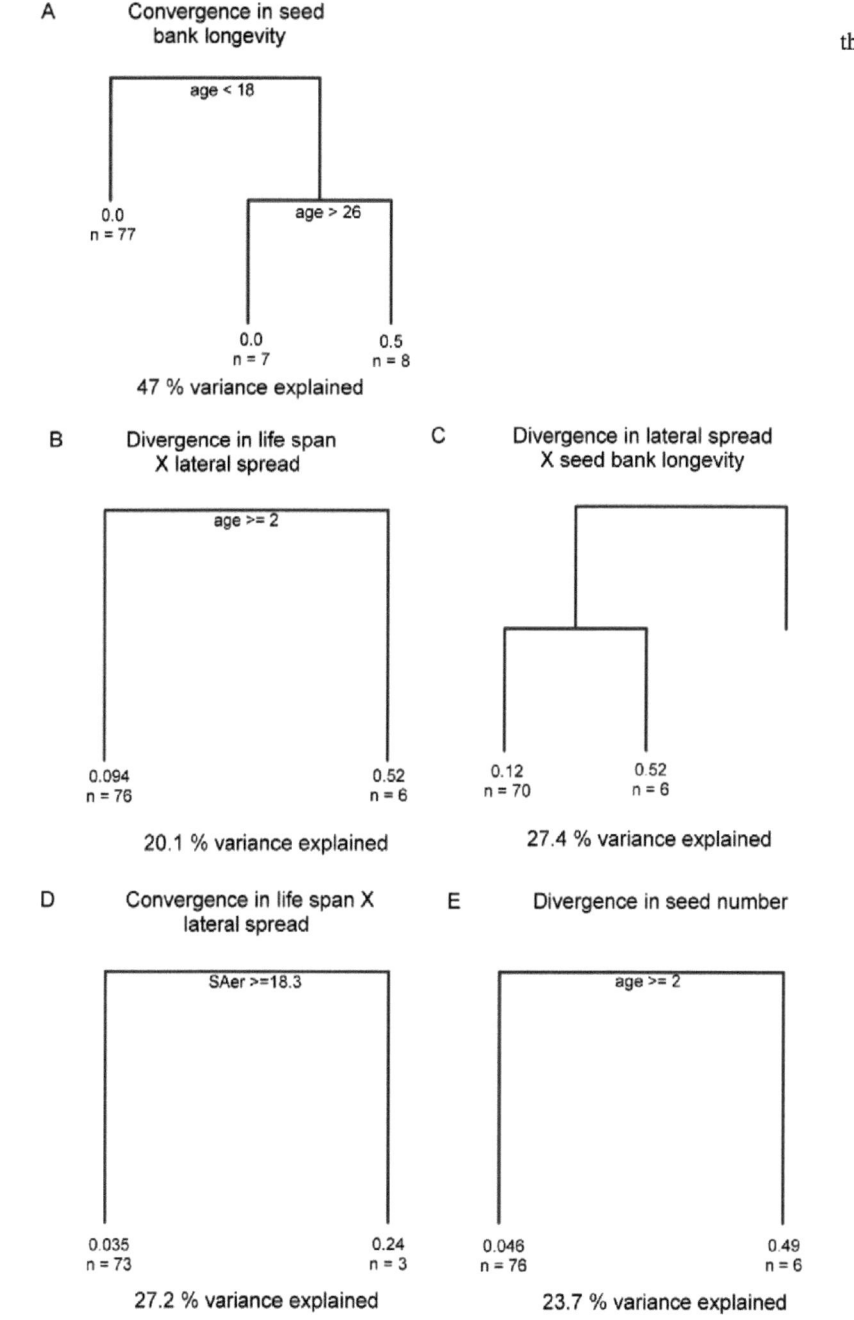

CHAPTER 6

neighborhood scale, however, four regression trees explained more than 20% of the variation. Convergence in the life span & lateral spread trait combination was more prevalent in plots with soil aeration <18.3% (n =3), compared to plots with reduced stress (n =73, Figure 17 D). The community weighted means of life span and lateral spread converged to intermediate life span (range: 2.23-2.40) and intermediate lateral spread (range: 1.26-1.51), compared to the remaining plots (ranges: 1.54-3.00 and 0.46-2.72, respectively).

Divergence in seed number and life span & lateral spread were largely restricted to early successional stages. Plots younger than 2 years (n =6) showed higher frequencies of divergence in seed number and life span & lateral spread than older plots (n =76, Figure 17 E and Figure 17 B). Increased frequencies of divergence in lateral spread & seed bank longevity were detected in plots with plot age <2 years or plot age >=27 years (n =6, respectively, Figure 17 C). At the intermediate plot age, only an average of 12% of the subplots showed divergence (n =70).

Table 13: Summary of the regression tree analysis results of convergence and divergence occurrences, observed at the community and neighborhood scales, in relationship to plot age (age) and soil aeration (SAer). Results in bold are referred to in the text. For each trait (combination) the location of the split (age in years, soil aeration in vol%) and the percentage of explained variance (in parentheses) are listed. If more than one significant split was obtained, the relationship with the variable that explained most of the variance is presented. NS designates non-significant splits. The function of interest for each trait (combination) is given by letters, which indicate the trait (combination) is relevant for dispersal capacity (D), competitive capacity (C), or stress tolerance (S). Trait combinations are denoted by a "&" between involved traits.

Functional trait	Function of interest	Community scale		Neighborhood scale	
		Convergence	Divergence	Convergence	Divergence
Seed number	D	Age <2 SAer >31.8 (13.9%)	NS	Age >=2 (3.3%)	**Age <2 (23.7%)**
Seed number/ terminal velocity	D	NS	NS	NS	NS
Seed number & SLA & life span	D	NS	NS	NS	NS
Canopy height	C	Age >0 (5.4%)	NS	SAer <34.6 (4.6%)	NS
Earliest flowering month	C	NS	NS	NS	NS
Flowering month & life span	C	NS	NS	SAer >=32.8 (8.4%)	NS
SLA	C/S	NS	NS	NS	NS
Lateral spread	C/S	NS	SAer >32.8 (17.1%)	NS	NS
Life span	C/S	NS	NS	SAer >=32.9 (11.9%)	NS
Life span & lateral spread	C/S	NS	NS	**SAer <18.3 (27.2%)**	**Age <2 (20.1%)**
Seed bank longevity	S	**18 <=Age <=26 (47.5 %)**	2 <= Age < 9 (12.5 %)	SAer <18.3 (4.5%)	NS
Lateral spread & seed bank longevity	S	NS	NS	NS	**Age <2 or Age >=27 (27.4%)**

CHAPTER 6

6.4 Discussion

In this study, we analyzed convergence and divergence for many traits at two different scales. However, the bulk of evidence was for trait-neutral assembly mechanisms. Not only did we find rare occurrences of convergence and divergence, their relationships with plot age and environmental stress were weak as evidenced by minimal variability explained in the regression trees.

6.4.1 Effects of dispersal filtering

Most functional patterns we elucidated were also not congruent with our hypotheses. During early succession, we expected convergence in dispersal traits but found divergence in seed number. This result can be explained as a consequence of an over-dimensioned increase in the occupied trait volume due to the prevalence of species with very high seed number (e.g. *Salix sp.*). This markedly outweighed the reduction in volume due to the absence of species generating very few seeds. Alternative data transformations to the log-transformed we applied might have altered these results (i.e. removed the divergence), but likely not yielded significant convergence.

Divergence in seed number was only loosely related to plot age. The other dispersal traits (seed number/terminal velocity and seed number & SLA & life span) did not show any strong functional patterns. We concluded that the dispersal filter did not make a notable contribution to the characterization of community assembly.

6.4.2 Effects of stress filtering

Our second hypothesis was that stress filtering is common along a successional gradient, which was not supported due to the rarity of significant functional patterns. In addition, the patterns were associated with plot age alone, which had no relationship to soil nutrient or water availability (Schadek *et al.* 2009).

We considered the effects of seed bank longevity and lateral spread as strategies to manage fluctuating environmental stress. A combination of these traits showed increased divergence in the youngest and the oldest plots at the neighborhood scale - where we had expected random patterns. Because the traits were not negatively correlated, higher FDis values may not necessarily suggest a trade-off between seed bank longevity and clonal propagation (Eriksson 1996; Ehrlén & van Groenendael 1998). Species coexistence with and without persistent seed banks, or with low and high lateral spread may induce divergence. Indeed, divergence was observed in plots dominated by three species groups: plants exhibiting transient seed banks and high lateral spread (e.g. *Corynephorus canescens*, *Agrostis capillaris*), species with persistent seed banks and no lateral spread (e.g. *Vulpia myurus*), and plants with persistent seed banks and clonal propagation (e.g. *Rumex acetosella*, *Carex hirta*). This observation, and in particular the last trait syndrome, suggests that investment in persistent seed banks as well as clonal propagation was favorable under some conditions in our study area - at least to the point of short-term persistent seed banks.

We found convergence in seed bank longevity at intermediate ages. However, this observation may be interpreted as a consequence of competitive exclusion rather than stress filtering, because shrub encroachment peaked at intermediate plot ages (data not shown). The absence of species forming short-term persistent seed banks under shrub canopies has previously been reported, and may be attributed to reduced regeneration success under strong competition (Milberg 1995; Davies & Waite 1998). This convergence was observed at the community scale, i.e. the 1 m² scale, where some large shrubs were apparently able to affect the functional characteristics of the entire community.

The present study provided evidence that stress filtering was restricted to divergence in the combination of seed bank longevity and lateral spread. However, the small percentage of variability explained (27.4%) in the regression tree suggested that stress filtering was not a viable predictor of functional community structure. This is consistent with the small degree of niche differentiation demonstrated by the OMI.

6.4.3 Effects of competition filtering

We assumed increasing convergence or divergence in traits associated with competitive ability at the neighborhood scale in later successional stages. However, convergence or divergence in canopy height or SLA, which are commonly associated with competitive ability, were not allied with plot age or environmental stress. Instead, life span & lateral spread yielded divergence in very young and convergence in the most fertile plots. The correlation between the two traits suggested that divergence in life span and lateral spread indicated coexistence of annuals and biennials, and perennials with high lateral spread in young plots. These results could be interpreted as evidence for limiting similarity. However it is not very plausible because limiting similarity implies strong competition, which is not likely at early successional stages with ample bare ground. Alternatively, we observed that perennials with high lateral spread, particularly *Festuca rubra*, were as equally fast at colonizing young plots as annuals. However, the perennial species required many years to establish a dense clonal network (as also noted by Grime 2002), and were unable to completely exclude plants with other trait expressions in later successional stages, particularly in stressful habitats. Competitive exclusion occurred only in the most fertile plots, where we detected significant convergence consistent with our assumption of increased competition filtering. Together with convergence in seed bank persistence at intermediate plot ages, this points to competitive exclusion as a mechanism generating convergence, rather than to limiting similarity which is associated with divergence.

Little empirical evidence for the limiting similarity theory is available. Fargione *et al.* (2003) provided the most compelling data, and suggested that inhibitory effects of established species were strongest for colonizers from the same functional guild. However, they also reported that a single, dominant group exerted the greatest negative effects. This is congruent with results of other studies that showed a functional group better adapted to local conditions can suppress other groups (e.g. Symstad 2000; Mwangi *et al.* 2007; Zavaleta & Hulvey 2007). However, competition filtering only occurred at the neighborhood scale and never scaled up to the community scale, which we expected at least for the later stages of succession. This may indicate that competition filtering was still operating and required extended time intervals to shape whole communities.

6.4.4 Evidence for trait-neutral assembly

Do we have to adopt the hypothesis of trait-neutral community assembly? From a methodological point of view, there are several sensitive decisions that might have affected the general outcome of our study. We selected eight traits and analyzed various combinations of these traits associated with dispersal, establishment and persistence (Weiher *et al.* 1999). However, due to logistic reasons we were unable to include below-ground traits, which have been suggested to be important in ascertaining competition in nutrient-poor ecosystems (Aerts 1999). Yet, plants distributed in our plots exhibited very low root densities (personal observation), inconsistent with the relevance of below-ground competition.

Another critical step is the choice of functional statistic (Lepš *et al.* 2006; Poos *et al.* 2009). FDis has only recently been proposed and rarely applied. Yet, FDis is related to Rao's quadratic entropy (Rao 1982) and is the multivariate analogue of the weighted mean absolute deviation (Laliberté & Legendre 2010), and both have successfully been applied in studies on functional diversity (e.g. Lepš *et al.* 2006; Kraft *et al.* 2008; de Bello *et al.* 2009). Similarly, in our study FDis was strongly correlated with functional richness as proposed by Villéger *et al.* (2008), which has shown suitable in plant and animal studies (e.g. Cornwell *et al.* 2006; Mason *et al.* 2008). Therefore, we have no reason to believe different results would have been generated if different statistics were applied.

The definition of the species pool is one of the most critical steps in null model approaches. Our species pool was based on annual botanical surveys in the study area since 2003 (see Schadek 2006) suggesting we did not overlook any taxa. However, we clearly recognize our conclusions are scale dependant, i.e. the species pool was limited to the study site which was rather uniform in terms of soil resources. If we increased the scale of the study area to the wet meadows of the adjacent marshland, strong environmental filtering would certainly have been revealed, as these species could not thrive on the dry, sandy landfills of the industrial site.

Based on the species pool of the study site, our results are highly in favor of our latter hypothesis. Community assembly appeared to be not the outcome of filter processes but of trait-neutral mechanisms. Neutral processes are an important force in many but not all terrestrial systems (see Hubbell 2001, and references therein). Yet, the degree neutral processes impact a community assembly is dependent on a number of community characteristics (Foster *et al.* 2004; Chase 2007), including successional progress. Although our time gradient spanned more than 40 years, competitive equilibrium may not yet have been established. Succession in resource-limited systems can encompass long periods of time (Schadek *et al.* 2009) extending the neutral processes phase. Indeed, several studies have empirically demonstrated that species composition patterns can remain largely neutral for up to 20 years of succession (Gitay & Wilson 1995; Cook *et al.* 2005; Holdaway & Sparrow 2006). Neutral mechanisms, such as arrival order, expose strong control on species composition during the first years of community assembly (Ejrnaes *et al.* 2006; Körner *et al.* 2008).

CHAPTER 6

Additionally, the dynamic character of our study area may have slowed succession. Generally, disturbances and recruitment limitations, among other causes, may prevent competitive hierarchies from operating (Huston 1999; Solé et al. 2004). Industrial areas are typically very dynamic in terms of habitat turnover rates and underlie a range of small-scale disturbances (Rebele 1994; Kattwinkel et al. 2009). Habitat turnover in urban environments constantly changes the spatial configuration of source habitats for dispersal and therefore may increase stochasticity in colonization events. Our measure of plot age captured habitat turnover effects. However, we have no insights into the relevance of small-scale disturbances that occurred at the study area before we initiated our work in 2003. These disturbances may have exerted long-term effects on the species composition in the plots, particularly by slowing down competitive exclusion.

Several studies have suggested neutral assembly processes at the taxonomic level (Cook et al. 2005; Ejrnaes et al. 2006; Körner et al. 2008), and our study found evidence that the initial phase of community assembly was largely neutral at a functional level. Our results support a trait-neutral view of community assembly in dynamic landscapes with rather homogeneous soil conditions, in which the relative importance of the spatiotemporal habitat configuration may override the relevance of filtering mechanisms.

Chapter 7

"Neutrality starts with assumptions that are clearly wrong, but produces patterns that match what we see in nature"

Jonathan Levine

7 Synthesis

It is a fundamental goal of ecology to understand the principles that govern biodiversity at multiple scales. While community ecology has long focused on assembly processes at local scales, metacommunity theory has shown that dispersal can drive biodiversity patterns not only at the landscape but also at the local scale. Four metacommunity paradigms have been suggested: patch dynamics, species sorting, mass effects, and neutral paradigm. The relevance of each depends on the importance of dispersal processes, niche differentiation along environmental gradients and ecological differences among species for structuring ecological communities. The applicability of metacommunity theory to natural plant communities is, however, hampered by the number of interacting species. The functional trait concept offers simplification by focussing on plant strategies rather than species identities. The concept assumes that narrow relationships exist between functional trait expressions and species ecological properties, which may be used to define functional groups. Contrasting trait expressions indicate contrasting ecological strategies and allow the classification of species into functional groups with different niches and dispersal capacities.

This work aimed at understanding the assembly of plant communities from a metacommunity perspective while adopting a functional trait approach. In a dynamic landscape, the relative importance of dispersal, stress and competition filtering for structuring brownfield communities was assessed by revealing relationships between functional traits and species response (i) to the spatiotemporal configuration of habitats, (ii) to abiotic stress, and (iii) to interspecific competition. Absence of such relationships was interpreted in favor of trait-neutral community assembly.

CHAPTER 7

Central questions were:

- What are the effects of the dispersal filter, the stress filter and the competition filter on the functional diversity of local plant communities and the constituent metacommunity?

    - What traits are responsive to what filter?
    - What filters are trait-neutral?

- What plant functional groups constitute the metacommunity?

- What may be concluded regarding the relevance of the four metacommunity paradigms for governing biodiversity in the study area?

In the following, the results of the conducted analyses are summarized and interpreted in the light of the four metacommunity paradigms. First, the functional relationships that were sustained are shown and the question is answered in how far filters provided important structuring forces during community assembly. Functional groups are derived from these relationships. Finally, these insights are interpreted in respect to the applicability of the four metacommunity paradigms to the study area.

7.1 Relationships between functional traits and assembly filters

7.1.1 Dispersal filtering

The applicability of metacommunity theory is conditioned on the demonstration of dispersal effects on biodiversity. Here, the role of dispersal filtering was assessed by the potential of the spatiotemporal habitat configuration (connectivity and age) to explain (i) the distribution of 52 plant species in brownfield patches (Chapter 4) and (ii) the distribution of trait expressions among locally coexisting species (Chapter 6). The neutrality of dispersal filtering was assessed by the identification of responsive traits out of a set of candidate traits typically associated with dispersal ability and competitive ability.

Two lines of evidence indicated that most species were dispersal limited. Firstly, connectivity was a significant predictor for more than half of the investigated species (Chapter 4) although the analysis focused on the 52 species most widespread in the study area. Given that lower occurrence frequencies imply larger distances between populations,

SYNTHESIS

even higher numbers of responsive species might be suspected if more rare plants had been included in the analysis. Secondly, even the best dispersers (i.e. plants with high seed number and low terminal velocity, see below) were missing in considerable fractions of patches. In 2006, only three species were recorded in more than 75% of patches (*Senecio inaequidens, Holcus lanatus, Agrostis capillaris*), which cannot be attributed to abiotic or biotic habitat unsuitability (Chapters 5 and 6). Instead, extraordinary events like strong winds or the passage of vehicles may have added stochasticity to the process of dispersal, which conforms to the view that dispersal limitation may act, at least partly, as a neutral force (Tackenberg *et al.* 2003; Soons *et al.* 2004).

Yet, trait differences between species responsive and non-responsive to connectivity indicated that dispersal filtering was not wholly neutral. Responsive species were characterized by a combination of high seed terminal velocity and low seed number. In other words, plants producing few seeds that took comparably short time to reach the ground (low terminal velocity) were more likely to be affected by the spatial configuration of habitat patches than other species.

But although a functional relationship with the dispersal filter was established, null model analysis demonstrated that the distribution of functional groups in the study area was largely unpredictable (Chapter 6). Divergence in seed number, observed in the youngest plots (age <2 years), may be interpreted as an overrepresentation of species generating many seeds in situations of maximum temporal isolation. But at average, this divergence was limited to 49% of the subplots defining one plot. Consequently, dispersal limitation appeared to be responsible for reduced abundances of species producing few seeds but was not strong enough to result in community-wide exclusion.

In fact, good and bad dispersers have repeatedly been observed to be equally fast in colonizing vacant sites (Miles 1987). To some extent, this may be attributed to species-specific connectivities. The habitat connectivity experienced by a given species in a distinct patch depends on the proximity and size of neighboring patches inhabited by the same species - but not by others. As a consequence, starting conditions for the colonization of a distinct habitat may vary enormously among species, even if they possess similar dispersal capacities. Thereby, the variation in species-specific habitat connectivities should increase with the level of dispersal: Every failure to colonize a new habitat impacts on the species connectivity and thus on its future colonization chances. This negative feedback may be particularly severe in dynamic landscapes where high habitat turnover rates bear the necessity of permanent dispersal. Like that, the dynamics of the study area might have added to the stochasticity naturally inherited in dispersal processes and promoted the generation of trait-neutral functional patterns.

7.1.2 Stress filtering

In the species sorting and the mass effects paradigm, abiotic habitat heterogeneity makes sure that each species finds "safe sites" within a landscape. This means, there are localities where the species is unaffected by competition because it is best adapted to the local environmental conditions and, accordingly, the "superior competitor". The present work addressed the relevance of environmental heterogeneity by investigating in how far stress imposed a constraint to the distribution of species within the study area. Therefore, an outlying mean analysis was calculated on locally measured soil parameters (Chapter 6). The analysis revealed that stress filtering played a minor role for the composition of local communities. Subsequent null model analysis confirmed this observation: Soil aeration as a measure of abiotic stress offered little predictability to the occurrences of convergence and divergence in traits associated with tolerance to stress.

Clearly, the irrelevance of stress filtering was limited to the extent of the study area and the associated species pool. Differences in the availability of resources may be important drivers of plant diversity (e.g. Hutchinson 1959; Whittaker 1975), including urban brownfield communities. Godefroid et al. (2007), as an example, demonstrated the potential of abiotic conditions to explain the composition and diversity of urban plant communities.

Evidence for stress filtering would probably have been obtained, had the moist grasslands surrounding the study area been included in the analysis. Since sandy filling materials were used consistently during development, the study area was expected to be rather homogeneous in respect to edaphic conditions. In fact, soil analyses confirmed minor differences among brownfields (Figure 18) and no marked relationship between the age gradient and soil resource availability (also see Schadek 2006).

Along the lines of almost identical levels of stress, no significant functional convergence and divergence in traits associated with stress tolerance was observed. Still, functional diversity within the species pool reflected the comparably nutrient poor and dry conditions of the study area. Soil aeration was generally high (15-38 volume %) pointing to comparably low water storage capacity and accordingly high variability of soil water and nutrients in upper soil layers. In such conditions, plants often show traits associated with tolerance or avoidance of water stress (Jentsch & Beyschlag 2003). Small canopy height and reduced leaf area have been suggested as indicators of tolerance to water stress. Similarly, short life spans and persistent seed banks enable annual plants to avoid the dry season by completing their life cycle before drought and to endure as seeds stored in the soil.

Many of the most prevalent species in the study area exhibited corresponding trait expressions. The perennial grasses *Festuca ovina, F. rubra* and *Corynephorus canescens*, which reduce the surface of their leaves by rolling or folding them, counted among the most abundant species of the study area. Regarding canopy height, only 16% of the species of the site were taller than 1 m. More than one third (36%) were annuals or biennials (Appendix 3). In summary, the functional diversity encountered in the species pool fitted well the expectation of stress tolerance traits. But because of an absence of a stress gradient, stress filtering was of no importance for explaining the functional composition of the brownfield communities in the study area.

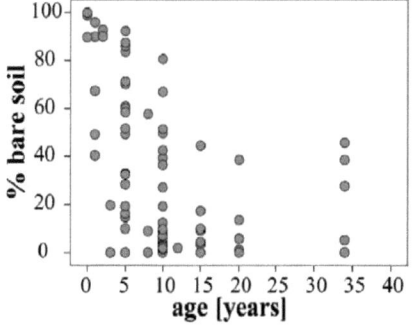
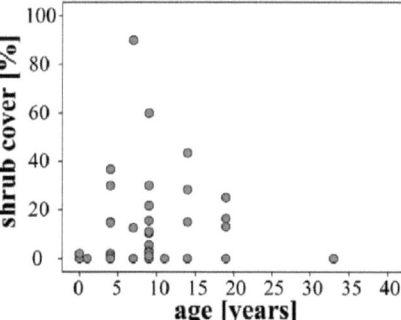

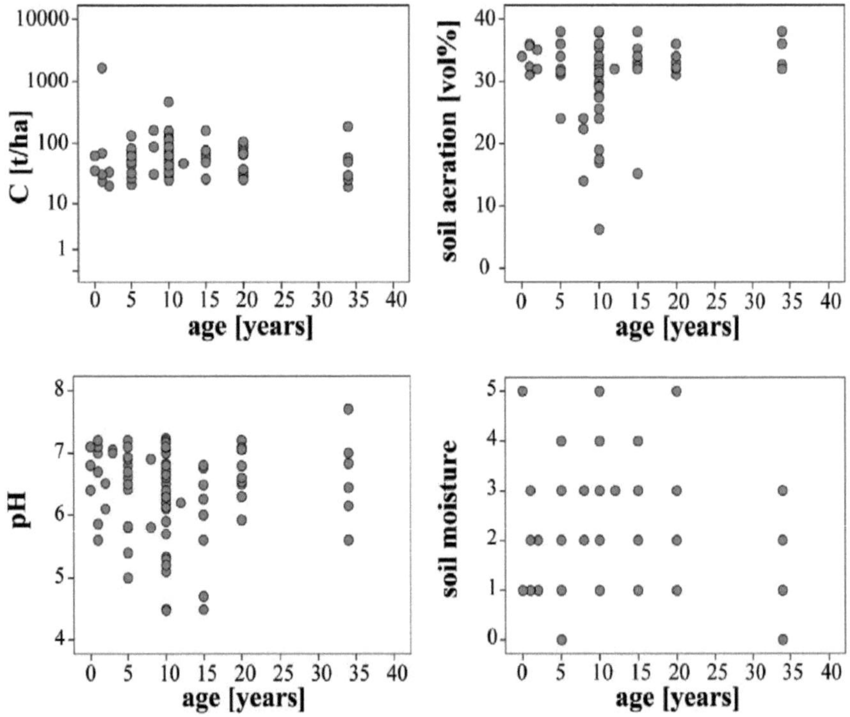

CHAPTER 7

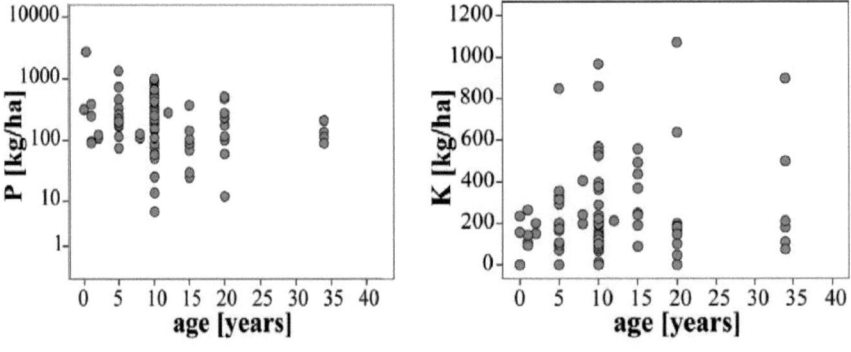

Figure 18: Plots of soil parameters against plot age. Presented are % bare soil, % shrub cover, carbon content (C), soil aeration, soil reaction (pH), soil moisture, phosphorous content (P), and potassium content (K) (from top left to bottom right). Note the log-scale of the y-axes of the plots of phosphorous, potassium and carbon content.

Figure 18: Extended.

7.1.3 Competition filtering

A particular aim of the present work was to explore the effects of resident species on colonizing species with a focus on the role of functional similarity for colonization success as well as for subsequent coexistence. Two analyses served this purpose. The role of competition for the establishment success of colonizers was addressed by seed bank analysis (Chapter 5). Long-term effects were examined by searching for shifts in the functional patterns during succession as a gradient of competitive pressure (Chapter 6).

Relevance of competition filtering
Null model and seed bank analysis provided contrasting results in respect to the relevance of competition filtering. Seed bank analysis demonstrated that colonization success was largely controlled by the availability of bare ground. Different mechanisms appeared to apply to plant groups with traits associated with high and low competitive response (i.e. weak and strong competitors), whereby short life span and high SLA indicated high competitive response. Weak competitors seemed to benefit from the specific microclimatic conditions for germination and establishment in young, largely unvegetated plots. By contrast, temporarily reduced competition after disturbance appeared to favor the establishment of strong competitors.

In the long run, these mechanisms should result in the exclusion of weak competitors from plots with low proportions of bare ground. Reduced functional diversity at increased plot age was, however, not supported by null model analysis. There was no evidence for significant functional convergence, neither in life span and/or SLA nor in any other of the investigated traits.

However, reduced frequencies of species with persistent seed banks were observed in the most fertile plots (i.e. low soil aeration, Chapter 6). In these plots, maximum shrub encroachment (*Salix spec.*, *Betula pendula*) was observed suggesting that competitive pressure peaked at maximum soil fertility rather than high plot age (Figure 19). Indeed, short-term persistent seed banks have already been associated with low regeneration success

CHAPTER 7

under shrub canopies (e.g. Milberg 1995; Davies & Waite 1998). Accordingly, the observation of convergence in seed bank persistence might be interpreted as the beginning of competitive exclusion. But it became significant in only three plots which leads to the conclusion that competition played no important role within the successional gradient of the study area.

Community-wide exclusion of weak competitors may take long time (Mal et al. 1997), and the more so under resource-poor conditions (Grime 2002; Foster et al. 2004). Low resource availability might have slowed down the pace of succession and limited the observation of competition filtering to the most fertile plots. But competitive exclusion may generally be a rather slow process. Modeling studies have shown that transient coexistence may be observed for thousands of generations even if one species is competitively dominant (Holt 2001). Assuming a generation time of two years competitive exclusion would take more than 500 years.

Moreover, competitive exclusion has been argued to be rarely absolute (Turnbull et al. 2005). Although establishment success may be considerably lowered, colonization still takes place. Failure to observe significant convergence or divergence following competition in natural plant communities may then be the rule rather than the exception. If so, competition filtering must be regarded as a largely neutral process constraining community assembly only to a minor degree.

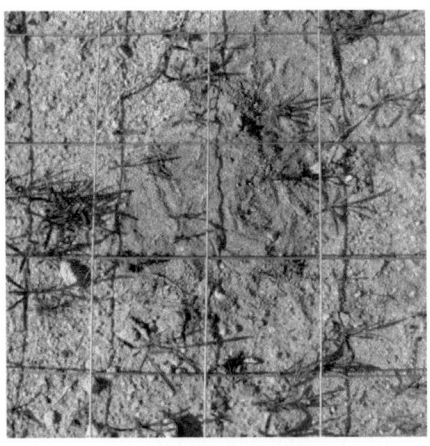
Age: 0 years; soil aeration: 36 vol%

Age: 7 years; soil aeration: 14 vol%

SYNTHESIS

 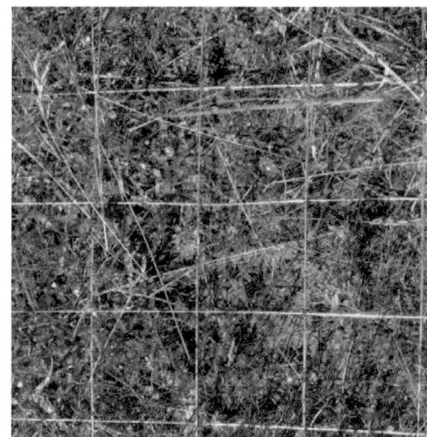

Age: 33 years; soil aeration: 33 vol%   Age: 7 years; soil aeration: 24 vol%

Figure 19: Examples of vegetation aspects differing in age and soil aeration.

CHAPTER 7

Lottery and dominance competition: Local persistence versus dispersal

The results of the seed bank analysis highlighted the relevance of bare soil, and consequently of lottery rather than dominance competition as a model of competition. Lottery competition assumes that competition works by excluding colonizers from localities (Holyoak *et al.* 2005), whereby vacant space is occupied on a "first-come, first-served" basis (Sale 1977; Warner & Chesson 1985). Contrarily, in the dominance competition model, strong competitors can displace weaker competitors from patches the latter already occupy (Hastings 1980) by lowering the amounts of available resources with negative consequences on the weaker competitor's growth, reproduction or other fitness components (Naeem *et al.* 2000; Prieur-Richard *et al.* 2000; Dukes 2001).

The conclusion of this work fits the row of studies questioning the potential of dominance competition to structure plant communities. In a review of competition studies, Goldberg & Barton (1992) concluded that resource partitioning was not an important mechanism of coexistence in plants - as a number of plant ecologists have argued before and after (e.g. Goldberg & Werner 1983; Shmida & Ellner 1984; Mahdi *et al.* 1989; Hubbell 2001). This may be particularly true when aiming at explaining vegetation shifts during succession. For instance, Halpern *et al.* (1997) observed that interspecific competition was definitely not responsible for the population decline of an early-successional plant species in a secondary succession. The general importance of small-scale disturbances for biodiversity patterns (e.g. Tilman 2004; Vandvik & Goldberg 2006; Eschtruth & Battles 2009), particularly during succession (Lavorel *et al.* 1994; Jensen & Meyer 2001) and in sand ecosystems (Jentsch & Beyschlag 2003), may be taken as further support for the relevance of lottery competition.

However, in the lottery competition model, all species possess equal probabilities to occupy vacant space (Sale 1977; Warner & Chesson 1985). But the results of this work indicated that establishment was not wholly neutral. Colonizers with high and low competitive response differed in their abilities to benefit from disturbance and bare soil (Chapter 5). This is in line with seed sowing studies reporting that establishment success was related to species traits (Gross & Werner 1982; Burke & Grime 1996; Roscher *et al.* 2009) making colonization dynamics akin to a "competitive weighted lottery" (Mouquet *et al.* 2004). Yet, the results obtained here suggested that a competitive weighted lottery may still produce neutral community patterns.

Role of functional resemblance

A main interest of this work was to understand the role of functional resemblance between colonizing and resident species during colonization as well as subsequent local coexistence. In the face of contrasting theoretical predictions, alternative hypotheses were formulated and tested. Based on the idea of competitive exclusion, increased colonization success was expected for colonizers that resembled the resident community in their functional traits resulting in trait convergence. According to the theory of limiting similarity, however, increased colonization success was expected for colonizers that were functionally different from established species generating divergence.

The idea of limiting similarity was not supported, neither by seed bank nor null model analysis. Functional divergence was assumed to be most likely under intense competition, i.e. under fertile conditions and/or late in succession. Yet, most traits did not show significant divergence as compared to null model expectations, and the rare observations showed no strong relationship with the levels of abiotic stress or the successional gradient. Similar insights were obtained by seed bank analysis. There was no significant relationship between colonizers and residents from the same functional group. Rather, the replacement of weak by stronger competitors added to the results of the null model analysis: Competitive exclusion rather than limiting similarity appeared to be the relevant mechanism behind competition filtering in the study area.

In fact, empirical evidence for limiting similarity is controversial - with at least three sources of confusion. Firstly, convergence and divergence are considered exclusive assembly mechanisms, i.e. either limiting similarity or competitive exclusion is expected. But often, both mechanisms are sustained in the same study. For instance, the idea of limiting similarity was supported by sowing and transplantation experiments. Resident species affected those colonizers strongest that belonged to the same functional group (Fargione *et al.* 2003; Mwangi *et al.* 2007). Yet, the same studies also noted that over all colonizers greatest inhibitory effects were exerted by a single, dominant group of residents. These observations provided evidence for both principles, limiting similarity and

CHAPTER 7

competitive exclusion, which begs the question: What mechanism will be decisive in the long run?

A second, yet related, source of confusion is the use of different measures of colonization success. Colonizer biomass is a useful and frequently used estimate (e.g. Fargione *et al.* 2003; Turnbull *et al.* 2005; Mwangi *et al.* 2007). Yet, the observation of reduced biomass does not imply absolute rejection. This means that, although colonizer biomass may be considerably lowered in functionally similar resident communities, colonization still takes place (Turnbull *et al.* 2005). Studies considering alternative measures of invasion success highlighted that different measures provided contrasting insights into the relevance of competition filtering and its functional relationships. For instance, Leishman (1999) observed that colonizer biomass was related to seed mass but survival to canopy height. Howard & Goldberg (2001) found that competition hampered seedling growth, whereas germination and seedling survival were not affected. These results stress the need for an extended monitoring of colonization experiments.

A third issue concerns the methodological difficulties in quantifying the effects of functional filtering. Navas & Violle (2009) predicted situations in which convergence and divergence were most likely. They suggested that divergence should be observable at intermediate levels of environmental stress and competition when coexistence of contrasting strategies is possible. This was not found in this work (Chapter 6). Yet, as also noted by Navas & Violle, the co-occurrence of strategies does not necessarily imply divergence, but simply multimodal trait distributions. If the peaks of such multimodal distributions occur at even distances, they may be captured by measures such as functional evenness (Villéger *et al.* 2008). However, this is not necessarily so. Co-existence of strategies is also addressed in the framework of guild proportionality, which predicts that the proportions of species belonging to a particular guild are more constant among communities than expected by chance (Wilson 1989). Guild proportionality has been demonstrated in several plant communities (e.g. Wilson & Roxburgh 1994; Gitay & Wilson 1995; Holdaway & Sparrow 2006). In such communities functional divergence and evenness measures could be validated. Yet, the link between guild proportionality and limiting similarity has so far not been made. This could be the first step to a quantification of the functional dissimilarity required for niche differentiation promoting the evaluation of the limiting similarity assembly rule.

Competition - a structuring force during community assembly?

Even if evidence for competition filtering is found in a community, inference on the relevance of competition as a structuring force is a somehow subjective decision. Competition can be regarded intense if it has major effects on various aspects of the performance of individuals or populations. The predictive value of competition for community composition and species coexistence within a community may still be small. What kind of evidence is required for formulating an assembly rule based on competition? Weiher & Keddy (1999) stated: *"Simply describing patterns is not the study of assembly rules"* They further proposed: *"An assembly rule specifies the values and domain of factors that either structure or constrain the properties of ecological assemblages."*

In the present work, functional resemblance between colonizing and resident species affected the establishment success of colonizers only to a minor degree - which extended to later stages of community formation: There was little evidence for increased competition filtering along the successional gradient. Overall, the relevance of competition for structuring communities must be regarded small, which is along the lines of experimental studies reviewed by Goldberg & Barton (1992). Almost all studies demonstrated competitive effects: The fitness of an individual or population was reduced when neighbors were abundant. Yet, there was little evidence that competition significantly influenced community structure.

Chapter 7

7.2 Plant functional groups in the study area

In this work, relationships of dispersal, stress and competition filtering with functional traits were determined to identify the prevailing assembly filters. Evidence for functional filtering was searched in a set of candidate traits assumed to be related to dispersal ability, stress and competition tolerance. Stress filtering was no important driver of community composition for the regarded species pool. By contrast, plant response to the remaining two filters was associated with functional traits. Dispersal filtering played a greater role for species exhibiting low seed number and high terminal velocity than for species with contrasting trait expressions. Competition filtering was sustained at the establishment stage. Strong competitors characterized by extended life span and low SLA largely constrained the establishment success of weaker competitors with contrasting trait expressions to young plots with ample bare ground. Inversely, the establishment of strong competitors was facilitated by the presence of weaker competitors.

These functional relationships allowed to discriminate plant functional groups with different responses to dispersal and competition filtering. Compared to "good dispersers", "bad dispersers" exhibited higher values of the quotient seed number/terminal velocity. "Strong competitors" showed longer life span and lower SLA than "weak competitors".

Did these two classifications operate independently from each other? Or did a trade-off among traits constrain the number of possible combinations of dispersal and competitive ability? Correlations of the quotient seed number/terminal velocity with life span and SLA were weak (Kendall`s $\tau$ / Spearman`s r <0.30, respectively) indicating that the correlative structure among traits posed no obstacle to the realization of any trait combinations in the species pool. Similarly, a principal component analysis (PCA) on the species in the study area, using their dispersal (seed number/terminal velocity) and competitive abilities (life span and SLA), showed that dispersal and competitive ability could be regarded as independent species characteristics (Figure 20).

SYNTHESIS

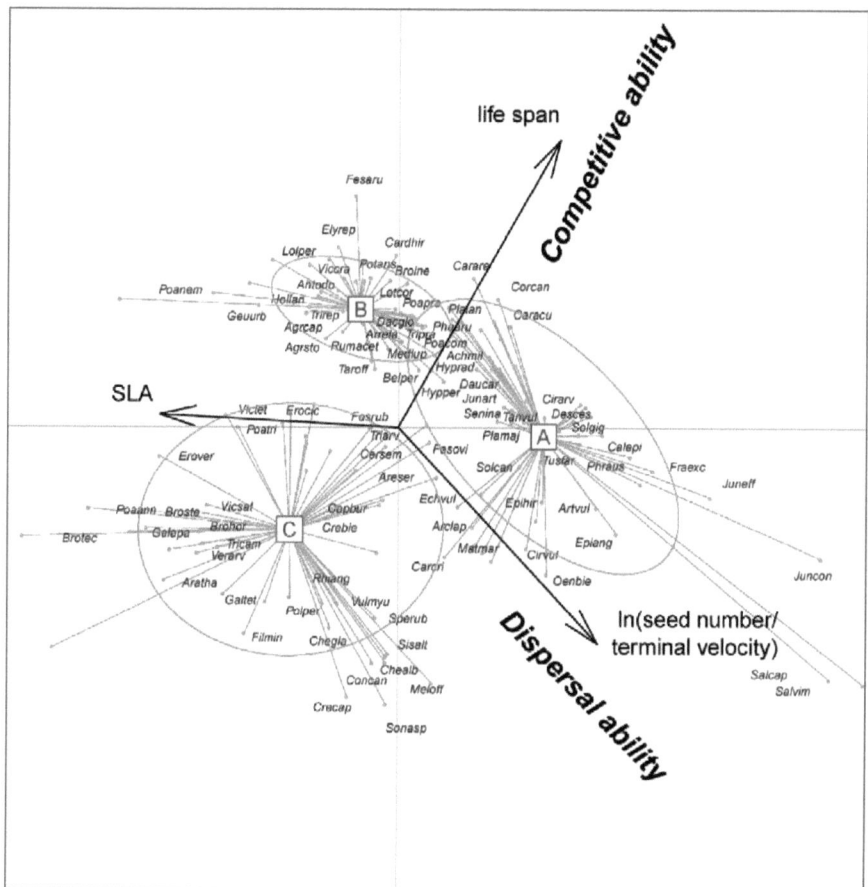

Figure 20: Ordination diagram (principal component analysis) of species of the species pool in the functional space span by traits associated with competitive ability (life span and specific leaf area: SLA) and dispersal ability (seed number/terminal velocity). Arrows represent functional traits. Ellipses indicate functional groups as obtained from hierarchical clustering of species scores on the first two ordination axes (Ward's method). The optimal number of groups was determined using Calinski criteria (Calinski & Harabasz 1974).

However, subsequent hierarchical clustering of the PCA-scores demonstrated that not all combinations of dispersal and competitive abilities were realized in the species pool. Three functional groups were obtained (Appendix 3):

- Group A: High competitive ability and high dispersal ability

- Group B: High competitive ability and low dispersal ability

- Group C: Low competitive ability and intermediate dispersal ability

While strong competitors covered the whole range of dispersal abilities (group A: high dispersal ability; group B: low dispersal ability), no combination of low competitive ability and high dispersal ability occurred. Weak competitors (group C) possessed higher dispersal ability than group B but lower than group A. This observation was contradictory to a competition-colonization trade-off. Weak competitors showed no superiority in dispersal ability. Consequently, they could not take advantage of a colonization deficit of strong competitors.

Another interesting finding was that one functional group combined high competitive and high dispersal abilities. But in spite of their favorable functional characteristics, species of this group were not the most prevalent in the study area: Out of the eight most abundant species in the plots in 2007 only *Senecio inaequidens* belonged to group A. This may be taken as another piece of evidence that dispersal as well as competition filtering played a minor role for community assembly in the study area.

7.3 Relevance of metacommunity paradigms in the study area

Based on the importances of dispersal, stress and competition filtering for governing biodiversity in the study area, the applicability of the four metacommunity paradigms may be discussed. The species sorting, mass effects, patch dynamics and neutral paradigm differ in the weights they confer to (i) dispersal, (ii) environmental heterogeneity, and (iii) ecological differences among species (see Table 14). The species sorting paradigm stresses the importance of variation of local abiotic conditions. If dispersal is sufficient to arrive in any locality of the landscape, species separate along gradients of resource availability according to their niche requirements. Environmental heterogeneity is also a central feature of the mass effects paradigm. But here, dispersal rates are so high that the fitness of local populations is directly affected by source sink dynamics, i.e. through the loss and gain of individuals through emigration and immigration. In the patch dynamics paradigm, niche

SYNTHESIS

differentiation does not rest on abiotic differences among local habitats. Instead, spatiotemporal niches are created by a trade-off between species dispersal and competitive abilities. The neutral paradigm differs from all other paradigms in that it assumes ecological equivalence of species. Biodiversity patterns are the result of stochastic extinctions and neutral dispersal limitation.

The results of this work underlined the absence of stress filtering. Accordingly, neither the species sorting paradigm nor the mass effects paradigms fitted the community dynamics of the study area. Yet, species sorting or mass effects might have played a role if the species pool had comprised species from adjacent habitat types. Within the framework of this work, however, the neutral paradigm and the patch dynamics paradigm provided better candidate metacommunity mechanisms which deserved to be discussed in more detail.

Table 14: Comparison of relationships between functional traits and dispersal, stress and competition filtering as expected by the four metacommunity paradigms (patch dynamics, neutral, mass effects, species sorting), and confirmed by the results of the present work (in boldface). For each filter, fills denote no relationship („-"), confirmed relationship („x"), and a relationship that may or may not be observed („(x)"). „x/-„ indicates functional relatoinshis with no consequences for the functional composition of local communities. More details are given in Chapter 2.4.

Dispersal filtering	Stress filtering	Competition filtering	Expected metacommunity paradigm
x	-	x	Patch dynamics
-	-	-	Neutral
(x) *	x	(x)	Mass effects
(x) **	x	(x)	Species sorting
x/-	-	x/-	**Observed relationships**

* high dispersal ability
** low dispersal ability

7.3.1 Patch dynamics paradigm

The applicability of the patch dynamics paradigm rests on three conditions:

- Species differ in the degree to which they are affected by dispersal limitation.

CHAPTER 7

- Some species are competitive superior, i.e. competition inhibits the long-term coexistence of weak and strong competitors within a locality.
- Colonization and competitive ability trade off, i.e. no species is superior in both aspects. Accordingly, newly created sites should be dominated by good dispersers but old sites by good competitors.

The results of this work suggested that the patch dynamics paradigm applied only to some extent. The first condition was met. There was dispersal limitation in the study area, and species differed in their response to the spatial configuration of habitats (i.e. connectivity). The second condition was not fulfilled. Although weak competitors established more successfully in the absence of residents, competitive exclusion was no structuring force. The third condition was also violated. Traits associated with competitive response were not responsive to dispersal limitation. Response to connectivity was associated with seed number and terminal velocity and not with indicators of competitive response: life span and SLA (Chapter 4). Moreover, the establishment success of strong competitors was not reduced in young plots (no significant relationship with bare ground, see Chapter 5), opposing the idea that young plots provided a spatiotemporal refuge for weak competitors. Finally, cluster analysis demonstrated that weak competitors were not superior in respect to dispersal. These observations contradicted a colonization-competition trade-off and hence to the patch dynamics paradigm as a mechanism of coexistence in the study area.

7.3.2 Neutral paradigm

The neutral paradigm regards trophically similar species equal. Neither the outcome of competition between a pair of species nor the arrival sequence of colonizers in habitats is predictable. Assuming that differences in competitive and dispersal abilities are manifested in functional traits, the neutral paradigm also denies relationships between functional traits and assembly filters. Accordingly, the paradigm is consistent with two observations:

- Biodiversity patterns do not differ significantly from random expectations.
- Relationships between functional traits and assembly filters are missing.

According to these conditions, evidence for and against the neutral paradigm was obtained in this work. Support for neutral mechanisms was provided by null model analysis (Chapter 6), which is commonly used to answer the question: How does a community look like if its assembly is governed by mere neutral processes? Null models generate neutral communities that can be compared to observed community patterns (e.g. Armbruster *et al.* 1994; Gotelli & McCabe 2002; Fridley *et al.* 2004; Jenkins 2006; Kraft *et al.* 2008). In this work, null model analysis demonstrated that, all along the successional gradient, functional community patterns were not different from neutral expectations. The number of plots with non-random functional patterns (i.e. with significant convergence or divergence) was generally low, irrespective of the considered trait and observation scale. However, the basic assumption of the neutral paradigm, the ecological equivalence of species, was opposed by evidence that dispersal limitation and response to competition during establishment were related to functional traits.

Evidence for and against neutral community assembly within the same metacommunity was also achieved by other plant ecological studies. Burns & Neufeld (2009) reinvestigated community turnover of island plant communities after a decade. Although having observed neutral community patterns in a previous study (Burns 2007), the comparison showed that extinction rates were correlated with stress tolerance traits - which is inconsistent with neutral theory. Similarly, Harpole & Tilman (2006) demonstrated in experimental grasslands that, although species abundances corresponded to neutral expectations, they were predictable from species competitive rankings.

One may argue that neutral community patterns alone are no definite evidence for the neutral paradigm. Metacommunity paradigms overlap in their predictions of community patterns (Bell 2005; Chase *et al.* 2005; Ruokolainen *et al.* 2009). As an example, community compositions that are consistent with neutral expectations are not exclusively attributable to the neutral paradigm but are also in line with the patch dynamics paradigm: Since there is no abiotic niche differentiation and hence no permanent refuge for weak competitors, local species composition and diversity may show great variability in time and space (Chase *et al.* 2005). Regarding the study area, however, the relevance of the patch dynamics paradigm was rejected because there was no support for a spatiotemporal niche differentiation among species (see section above).

If the patch dynamics paradigm has to be denied as an alternative source of neutral community patterns, how does the observation of functional relationships fit the neutral paradigm? In fact, the neutral model in its strict version is rarely applicable to natural communities. Not even proponents of neutral theory contradict the existence of ecological differences among species (e.g. Bell 2001; Hubbell 2001). Rather, their explanatory value is questioned. To say it in the words of Graham Bell (in Gewin 2006, p. 1308): *"Certainly selective differences are operating, but it is interesting to see whether those differences are large enough to overcome stochastic (or random) forces"*. Consequently, to reject the neutral paradigm it has to be demonstrated that neutral mechanisms are inferior to non-neutral mechanisms in explaining biodiversity patterns.

The results obtained here suggested that the ability of dispersal as well as competition filtering to structure whole communities was indeed limited. Dispersal filtering did not generate significant functional patterns. Although this may to some extent owe to the fact that low dispersal capacities were compensated by habitat connectivities (see Chapter 7.1), the low predictive value of habitat connectivity suggested that, even at a species level, dispersal filtering added little predictability to the distribution of plant species in the study area. Similarly, null model analysis indicated that, although competitive filtering at the establishment stage was supported, it was not strong enough to exclude weak competitors from any community along the successional gradient considered. As a conclusion, dispersal and competition filtering appeared to act as largely neutral forces in the study area.

7.3.3 Hybrid metacommunity models

Out of the four main paradigms of metacommunity theory, evidence of this work was strongly in favor of the neutral paradigm. For the sake of completeness, it should however be mentioned that hybrid models that combine the patch dynamics paradigm and neutral mechanisms might be applicable to the study area, too.

Coexistence under the patch dynamics paradigm is not strictly contingent on the presence of a competition colonization trade-off. Modeling studies have shown that next to a trade-off other mechanisms can create spatiotemporal niches allowing coexistence at the regional scale. Such mechanisms can either emphasize differences between species or be neutral. A non-neutral mechanism is for instance asymmetric competition: Strong competitors can establish everywhere but competitive weak seedlings only in the absence of strong competitors (Amarasekare 2003). This mechanism, however, conflicted with the results of the seed bank analysis. Strong competitors established most successfully after disturbance, indirectly suggesting that resident species reduced their colonization success.

Neutral additional mechanisms have, for instance, been described by Calcagno et al. (2006). They showed that coexistence in the framework of lottery competition is possible, if dispersal abilities are not too low, and if there is an upper limit to the competitive dissimilarity between species. However, although the authors suggested that both conditions may be frequently met in nature, more detailed data is required to answer the relevance of such a mechanism in the study area.

Another example for an additional neutral mechanism was presented by Kisdi & Geritz (2003). They introduced fine-scale heterogeneity into their model and showed that patch dynamics can promote coexistence, if the number of seeds in a locality – the candidates for establishment - is subject to demographic stochasticity. This model might have applied to the study area. Strong variability in seed bank abundances has often been observed in nature (Jentsch & Beyschlag 2003; Fenner & Thompson 2004) and was also sustained in the study area (see Chapter 5). Such variability may result from spatial heterogeneity in seed pressure. In local seed banks, those species might prevail that produce high quantities of seeds in adjacent localities. In the study area, community composition in newly created patches appeared indeed to reflect species dominating in neighboring patches that had shed their seeds during the time of patch creation (personal observation).

These examples emphasize that hybrid models provide important enlargements of the four main metacommunity paradigms. Yet, consideration of all these models increases significantly the effort to identify the main drivers of community assembly. For the purpose of the present work, a focus at the four main paradigms may therefore be preferable.

7.4 Predicting the relevance of metacommunity paradigms in plant communities

The results of this work were consistent with the neutral paradigm although the species sorting and mass effects paradigms have been suggested to be the most common metacommunity paradigms in nature (Leibold et al. 2004; Holyoak et al. 2005). In meta-analyses of Cottenie (2005) and Leibold & Mikkelson (2002) the majority of data sets was structured by species sorting processes. However, the results of meta-analyses may be misleading: Most ecological data sets may be assumed to have been sampled along abiotic gradients to be able to explain as much community variability as possible. In the absence of

CHAPTER 7

such a gradient, abiotic factors, and hence the species sorting paradigm, will prove to be little influential on structuring plant communities. Then, the neutral and the patch dynamics paradigm may act as mechanisms of coexistence as observed in the present work.

Indeed, patch dynamics models have a long-standing tradition in plant ecology (e.g. Watt 1947; Tilman *et al.* 1997). They share many aspects with other plant ecological concepts including the idea of fugitive species (Hutchinson 1951; Horn & MacArthur 1972), Whittaker's successional niche concept (1969), and the regeneration niche concept (Grubb 1977; Pacala & Rees 1998). All these models have in common that weak competitors avoid area-wide extinction by taking advantage of spatiotemporal niche opportunities. Patch dynamics have been deemed responsible to explain species coexistence in savanna ecosystems (Meyer *et al.* 2009), in perennial grasslands (Questad & Foster 2008) and in rangeland ecosystems (Kleyer *et al.* 2007).

Yet, few of these studies attempted to assess the relevance of neutral processes as an alternative mechanism of coexistence. Still, the stochasticity of colonization sequences has been noted repeatedly including the colonization sequences of functional groups (Miles 1987; del Moral & Grishin 1999; Frelich & Reich 1999; Lawrence & Ripple 2000).

Only recently, studies explicitly started to address the relative contribution of neutral and deterministic (i.e. non-neutral) processes to explain biodiversity patterns. Along the lines of the conclusion obtained here, such studies observed that neutral and deterministic processes often controlled biodiversity simultaneously (Wootton 2005; Ellis *et al.* 2006; Driscoll & Lindenmayer 2009). For instance, Perry *et al.* (2009) fitted a neutral model to abundance data of Australian shrubland communities. They found that, although species turnover was attributable to differences among species in respect to dispersal and establishment success, species abundance distributions conformed to neutral expectations. Similar insights were provided by sowing experiments. Ejrnaes *et al.* (2006) observed that the competitive ability of colonizers controlled species richness, although arrival sequence was the best predictor of species abundances. Questad & Foster (2008) found that the composition of sown communities was strongly influenced by neutral as well as non-neutral forces.

Such observations are hardly surprising. *"(I)t is unlikely that all the species that interact in a given set of real metacommunities will uniformly conform to any one of these (four metacommunity) perspectives"* (Leibold *et al.* 2004, p. 608). In other words, the relevance of environmental heterogeneity is not only an attribute of the landscape but is also contingent on the niche widths of inhabiting species. Kolasa & Romanuk (2005) suggested a relationship between spatial resolution of environmental heterogeneity and habitat specialization. The more specialized the species, the greater the habitat heterogeneity the species perceives. Consequently, the paradigms species sorting and mass effects may be adequate to describe the interactions between species groups with different habitat perceptions, whereas the neutral and the patch dynamics paradigm may be favorable for species with comparable perceptions.

Similarly, differences in dispersal abilities imply that species perceive habitat isolation in unique ways (e.g. Dupré & Ehrlén 2002; Kolb & Diekmann 2005; Endels *et al.* 2007). The same habitat configuration can provide an efficient dispersal barrier for the one species but may be highly permeable for another. Indeed, in the meta-analysis of Cottenie (2005), the relevance of dispersal-based metacommunity paradigms (patch dynamics and neutral paradigm) was lower for taxa with passive dispersal (i.e. requiring a vector) amongst which counted terrestrial plants.

Accordingly, a shift in the relevance of metacommunity paradigms along scales may be hypothesized. Species sorting should be more important at coarse scales, which are likely to incorporate vast geographic or altitudinal gradients, than at fine scales. In fact, in studies investigating biodiversity change at coarse scales (i.e. > ~100 km), abiotic parameters and dispersal limitation have been suggested as main drivers of biodiversity (e.g. Borcard *et al.* 1992; Gilbert & Lechowicz 2004; Perry *et al.* 2009; Svenning *et al.* 2009). Thereby, the magnitude of dispersal limitation, i.e. the fraction of empty but potentially suitable habitats, increased with larger geographic distances (Freestone & Inouye 2006) and decreased with the dispersal capacities of species (Ozinga *et al.* 2005; Schurr *et al.* 2007).

By contrast, neutral mechanisms may be more important at finer scales. In a forest ecosystem, Girdler & Barrie (2008) reported that neutral forces indicated by increased amounts of unexplained variability in community composition were stronger at finer scales than at coarse scales. This relationship may, however, only hold if differences in scale are not overridden by differences in the dispersal capacities of species. In the previously mentioned meta-analysis of Cottenie (2005), dispersal ability was a better predictor of the

CHAPTER 7

relative importance of niche-based (i.e. species sorting and mass effects) and dispersal-based (patch dynamics and neutral paradigm) metacommunity paradigms than scale.

The relevance of the neutral paradigm stated in this work fitted the above observations. Presumably high dispersal capacities of ruderal species contrasted a comparably fine scale making the study area "prone" to dispersal-based paradigms.

7.5 A glance at methods

This work showed that the neutral paradigm may offer a thorough understanding of the first stages of plant community assembly, at least in dynamic systems such as the study area. Such insights are rare because few metacommunity studies have dealt with natural plant communities and even few were conducted in dynamic landscapes. Knowledge of the relevance of dispersal processes for plant diversity in dynamic landscapes is largely constrained to modeling studies (e.g. Keymer *et al.* 2000; Johst *et al.* 2002). Empirical studies are urgently required to verify the realism of assumptions and the applicability of their predictions to nature. The results of this work were retrieved under natural conditions. The spatiotemporal habitat dynamics, as well as taxonomic and functional diversity were borrowed from a real landscape. Naturally, this realism came at some cost.

Trait-filter assumptions

In this work, evidence for and against dispersal, stress and competition filtering of functional traits was used to determine the relevance of the four metacommunity paradigms. This approach rests on the same idea as variance partitioning (see Chapter 2.1), but may be regarded superior in that it sheds light on the relevance of species ecological differences. It was this functional approach that enabled the investigation of non-neutral filter mechanisms in the highly diverse plant communities of the study area.

Though, the existence of ecological differences among species in their abilities to disperse and to cope with stress and competition was based on assumed relationships between these abilities and functional traits. But responsiveness of functional traits may be the result of the correlative structure among functional attributes or phylogenetic relations among species (Ackerly & Reich 1999; Tremlová & Münzbergová 2007). A verification of the causality of these relationships was, however, beyond the scope of this work. Detailed knowledge on the actual dispersal rates and distances as well as competitive rankings of all involved species would have been required. Regarding the high number of encountered species (255 species made up the species pool) this was logistically impossible. Merely determining competitive rankings in diverse plant communities has been called *"nightmarish"* (Goldberg 1996, p. 1381). However, the causality of the assumed functional relationships has been sustained by a large number of studies including modeling (Tackenberg *et al.* 2003; Warren & Topping 2004; Bossuyt & Honnay 2006), experimental (Howard & Goldberg 2001; Liancourt *et al.* 2009) and field studies (Suding *et al.* 2003; Verheyen *et al.* 2004; Wright *et al.* 2004b; Ozinga *et al.* 2005).

Separation of paradigms

Alternatively to the determination of trait-filter relationships, key predictions of the four paradigms (see Table 15) could have been contrasted to reveal the prevalent metacommunity mechanisms (e.g. Ellis *et al.* 2006; Driscoll & Lindenmayer 2009; Pandit *et al.* 2009). But this approach is often of limited use in empirical studies like the present one.

For instance, to separate the patch dynamics from the neutral paradigm, information on long-term community dynamics is required, including rates of speciation, migration and death, as well as population sizes of all species composing the metacommunity. Such information is almost impossible to obtain for species-rich communities leaving some ecologists to question the direct testability of neutral theory (e.g. Enquist *et al.* 2002; Chase *et al.* 2005; Cottenie 2005; Adler *et al.* 2007).

Another debated prediction used to disentangle the four metacommunity paradigms concerns community change after disturbance. Chase *et al.* (2005) stated that there are no consistent statements that allow to disentangling the patch dynamics and the neutral paradigm. Indeed, in a review of Mackey & Currie (2001), more than one third of studies found no effects of disturbance on species richness at all.

Other predictions may allow a separation of the patch dynamics and the neutral paradigm under distinct conditions, but did not conform to the design of this work. Predictions, referring to regional metacommunity characteristics (e.g. dispersal effects on regional diversity), as an example, rest on a comparison of several metacommunities. Consequently, they are not testable by studies dealing with a single system of connected habitat patches.

CHAPTER 7

Similarly, a comparison of local and regional diversity along dispersal gradients would probably not have distinguished the two paradigms in this work. There was no single dispersal (or connectivity) gradient describing the degree of habitat configuration for all species. Habitat connectivity was a species- (and patch-) specific attribute. Abundance distributions, finally, allow a separation only if dispersal is global (Chase *et al.* 2005) – which is probably rarely the case in fragmented landscapes.

The present work thrived for a more holistic view of community assembly to increase the reliability of observations (see Bell 2005). Community patterns were investigated at different scales (patch, plot and subplot) - and with different foci. Two analyses aimed explicitly to elucidate the neutrality of dispersal and competition filtering (Chapters 4, 5). This approach resulted in the important insight that, although community compositions did not differ from neutral expectations, dispersal and competition filtering still involved functional traits.

SYNTHESIS

Table 15: Summary of predictions of the four metacommunity paradigms (neutral, patch dynamics, species sorting, mass effects). Note: Predictions without superscripts are speculation not yet backed up by specific theory. By Chase et al. (2005).

Effect	Model prediction			
	Neutral	Patch dynamics	Species sorting	Mass effects
Overall local diversity	Extinction and colonization balance [a]	Extinction and colonization balance	Depends on species interactions	Depends on species interactions and balance between extinction and colonization
Overall regional diversity	Extinction and speciation balance [a]	Depends on competition-colonization trade-off	Same as above and degree of habitat heterogeneity	Same as above and degree of habitat heterogeneity
Relative species abundance	Zero-sum multinomial (skewed toward rare species) [a]	Variable depending on level of migration and degree of interaction [c]	Variable depending on environmental conditions	Variable depending on level of migration [d]
Dispersal effects: local diversity	Increase	Hump-shaped [d]	No effect	Hump-shaped [e]
Dispersal effects: regional diversity	Decrease	Decrease	No effect	Decrease [e]
Dispersal effects: beta-diversity	Decrease [a,b]	Global: no effect Local: decrease	No effect	Global: decrease [e] Local: decrease
Local disturbance	Return immediately	Unpredictable	Return immediately	Return following succession
Regional disturbance	Random walk	Return following succession	Return immediately	Return following succession
Temporal variation: local	Variable	Variable	Static unless environment changes	Static unless environment changes
Temporal variation: regional	Variable	Static unless environment changes	Same as above	Same as above

[a] Hubbell (2001), Bell (2001); [b] Chave & Leigh (2002); [c] Chave et al. (2002); [d] Mouquet et al. (2002); [e] Mouquet & Loreau (2003)

CHAPTER 7

7.6 Future research needs

This work has touched a number of ecological concepts concerned with the principles that govern biodiversity. While some of these have a long-standing history (e.g. the limiting similarity theory of MacArthur & Levins 1967), metacommunity theory and neutral theory have sparked ecological interest more recently (e.g. Hubbell 2001; Leibold *et al.* 2004), and much is still to be learned. During the conduct of this work, some insights related to metacommunity theory and the functional trait approach became apparent that deserve the attention of future studies.

Time for experiments

The role of metacommunity mechanisms in plant communities is still little explored. Although there is growing consensus that dispersal processes may play a fundamental role for community properties, their implications are rarely addressed. For instance, metacommunity studies dealing with small organisms (microbes, butterflies, zooplankton) have demonstrated that subtle changes of habitat configuration or the presence of predators may have profound consequences for the stability of the whole community (e.g. de Meester *et al.* 2007; Louette & De Meester 2007). Certainly, these experiments were facilitated by the short regeneration cycles of organisms and the ease to manipulate attributes of their habitats.

Still, experiments might be feasible for herbaceous plant communities, too. Weed communities are often composed of comparably short-lived species with the ability to colonize spontaneously. This may provide good starting conditions to survey metacommunity dynamics across larger time scales. Certainly, experiments in real landscapes, as the brownfield sites of this work, may be regarded difficult. As many urban systems, the investigated brownfields were subject to frequent industrial and recreational uses turning already the maintenance of permanent plots over 3 years into a challenge (46% of plots underwent some type of disturbance). But an investigation of metacommunity dynamics under more controlled conditions might not only promote our understanding of the long-term effects of dispersal during community assembly. Experimental approaches further allow manipulating environmental conditions (e.g. spatial habitat configuration, abiotic habitat heterogeneity, competitive pressure) and studying consequences for the dynamics of plant metacommunities.

SYNTHESIS

Predicting the relevance of metacommunity paradigms

Habitat isolation and environmental heterogeneity are no mere landscape attributes. Plants perceive spatial and temporal distances as well as abiotic conditions in different ways. In a previous chapter (Chapter 7.3), hypotheses were derived how interactions of study scale, environmental heterogeneity, dispersal ability, and niche width may be related to the relevance of the four metacommunity paradigms. This idea is not new (e.g. Leibold *et al.* 2004) but rarely has it been subject to empirical or experimental validations (but Freestone & Inouye 2006; Girdler & Barrie 2008). Testing the relationship between the relevance of metacommunity paradigms, species properties at different scales and degrees of habitat heterogeneity might, however, decidedly elucidate the situations in that community assembly is controlled by deterministic or neutral forces.

Such tests may also serve to explore the neutrality of single mechanisms that show neutral as well as non-neutral characteristics. For instance, evidence of dispersal limitation has often been interpreted in favor of the neutral theory although differences among species in their abilities to disperse are obvious (see Clark 2008). The interaction of disturbance timing and time of seed shedding may be considered as another example of a mechanism that has been categorized neutral but is associated with species traits (here: time of seed release). Identifying the situations in that such mechanisms should be considered neutral or deterministic might considerably contribute to enhance the predictability of biodiversity.

Understanding the role of biotic interactions during assembly

The role of biotic interactions for local and regional plant diversiy is only poorly understood. While investigation of pairwise species-interactions is of limited use to describe species-rich plant communities, functional approaches have yielded ambivalent results (e.g. Wilson & Gitay 1995; Watkins & Wilson 2003; Stubbs & Wilson 2004).

Functional approaches may have suffered from the neglectance of long-established plant ecological knowledge. For instance, the relevance of limiting similarity during colonization has received the strongest support from studies investigating competition effects between legumes, grasses and forbs (e.g. Fargione *et al.* 2003; Mwangi *et al.* 2007). These species groups have long been accepted to follow different strategies of resource use. Mutualistic relationships with nitrogen fixing bacteria allow legumes to cope with low nitrogen levels in

CHAPTER 7

unique ways. Such characteristics have however rarely been related to functional traits. Yet, understanding how long-established plant strategies are reflected in e.g. functional evenness, divergence and dispersion (Villéger *et al.*) might decidedly promote the validation of the limiting similarity idea and its value to act as an assembly rule for plant communities.

Moreover, most metacommunity models focus on negative biotic interactions (i.e. competition) and their significance during community assembly (but Klausmeier 2001). However, positive interactions, too, may play an important role during the formation of plant communities (e.g. the facilitation model of Connell & Slatyer 1977). And there is growing evidence that facilitation is not constrained to stressful conditions (Goldberg & Barton 1992; Bruno *et al.* 2003; Brooker 2006). Therefore, the role of facilitative relationships during community assembly deserves more attention in metacommunity studies.

7.7 Outlook

Are there general rules that allow predicting communities in a given place in space and time? In the present work, an attempt was made to understand community assembly as a functional filtering process, during which the dispersal filter, the stress filter and the competition filter consecutively constrain functional diversity. Once the dispersal filter has defined a pool of locally available species, local abiotic and biotic factors should explain the distinct functional composition of a community. However, the present work showed that community predictability may remain low even if dispersal and competition filter each influence distinct functional groups in specific ways. Instead, community assembly appeared to be largely driven by trait-neutral processes.

From a metacommunity perspective, this conclusions points to the neutral paradigm, which has implications for the management and conservation of biodiversity (Loreau *et al.* 2003a; Mouquet & Loreau 2003). While the species sorting and the mass effects paradigms highlight the role of habitat heterogeneity, the neutral paradigm (and the patch dynamics paradigm) stresses the importance of dispersal processes. Dispersal is required to compensate for local extinctions and community recovery after disturbance (Chase *et al.* 2005). In such situations, conservation management should aim at ensuring migration among habitat patches by preserving the characteristics of their spatiotemporal patch configuration. Indeed, in the case of the study area, this emphasis of dispersal processes is

SYNTHESIS

well in agreement with empirically evolved management recommendations (Jentsch & Beyschlag 2003; Schadek 2006; Kattwinkel *et al.* 2009).

Similarly, extinction sequences following habitat loss or fragmentation are unpredictable under the neutral paradigm, i.e. decreased habitat connectivities do not necessarily disadvantage populations of species with low dispersal capacities. The results of this work particularly highlighted species-specific notions of habitat connectivity as an important trait-neutral factor. Even the survival of good dispersers may be at risk if those habitats become unavailable that act as a source or stepstone. Consequently, to maintain (meta-) populations of a given focal species in such a landscape, nature conservation must identify and conserve those habitat patches that are crucial for its persistence.

However, the results of this work also suggested that traits may become more relevant for predicting extinction sequences if patch turnover rates change. As the establishment of weak competitors appeared to be hampered by closed vegetation canopies, they may be particularly prone to extinction if suitable sites for recruitment are not within (spatiotemporal) reach. Regular patch turnover may be required to guarantee suitable establishment sites and inhibit the encroachment of shrubs in the long run. By contrast, strong competitors may be less endangered as they are well able to establish into extant communities. If so, competitive superiority rather than dispersal superiority might promote persistence under reduced habitat dynamics. In the end, however, only long-term observation of population dynamics under different patch turnover rates can answer the question in what situations extinction sequences in dynamic landscapes are contingent on traits or conform to neutral expectations.

Accepting the neutral paradigm as the prevalent metacommunity paradigm also has consequences in respect to the long-term provision of ecosystem functions (Loreau *et al.* 2003b). The biodiversity insurance hypothesis states that migration between local communities provides insurance for ecosystem functioning because species unable to cope with altered environmental conditions can be replaced by better adapted ones (Loreau *et al.* 2003a). Knowledge of the prevalent metacommunity paradigm may allow to predict which species cannot be replaced, as well as the ecosystem functions related to the loss of that species. Under the neutral paradigm, however, the loss of any particular species will have no effect on ecosystem properties (Bell 2005). Yet, little can be said upon the predictability of

CHAPTER 7

ecosystem properties in the study area. Neutral community assembly does not imply neutrality in respect to the provision of ecosystem services. Although the assembly of the brownfield communities was largely neutral, species in the regional pool (including annual and perennial forbs, grasses, shrubs and trees) were not necessarily equal in terms of water and carbon storage, litter decomposability, biomass production, and so forth.

In summary, the present work has shown that even if the functional trait approach offers limited predictability of plant community assembly, in concert with metacommunity theory it may contribute significantly to reveal the most important mechanisms during community assembly. The integrative approach showed that the first years of community assembly may be largely neutral, at least in a dynamic landscape. Accordingly, predicting the consequences of environmental change on biodiversity requires more complex information than just knowledge of species traits and local habitat characteristics (Clark 2008). This result should be understood as an urgent hint that stochastic forces need to be implemented when one aims at understanding and preserving biodiversity in dynamic landscapes.

8 Appendix

Appendix 1: List of figures.

Figure 1: Schematic representation of the four paradigms of metacommunity theory for two competing species with populations A and B. 12

Figure 2: Separation of the four metacommunity paradigms (SS: species sorting; ME: mass effects; PD: patch dynamics; NM: neutral) along gradients of dispersal and environmental heterogeneity. 16

Figure 3: Schematic illustration of community assembly as a hierarchical process..... 22

Figure 4: Illustration of the distribution of functional trait expressions following the filtering of traits from a given species pool. 27

Figure 5: Sequence of aerial photographs of a subarea of the study area from 1974, 1987 and 2002. 39

Figure 6: Overview of observation scales considered to assess the effects of dispersal (patch scale), stress (plot scale) and competition filtering (subplot scale). 41

Figure 7: Overview of the study site, the subdivision into patches and the location of plots. 42

Figure 8: Comparison of functional divergence and functional convergence. 47

Figure 9: Frequency of connectivity weights for the studied species. 61

Figure 10: Relations between selected plant functional traits (terminal velocity, log seed number, log seed mass, log (seed number/terminal velocity)), and species groups based on their response to connectivity. 64

Figure 11: Overview of trait expressions of the functional groups of three different group definitions (groupings: "Establishment", "Resource capture", "Strategy"). 84

Figure 12: Outlying mean index analysis (OMI) of 134 plant species observed in the study area on local environmental variables. 88

Figure 13: Flowchart of the computational steps performed for each trait and trait combination to assess significant functional convergence and divergence, at the community and at the neighborhood scale. 110

Figure 14: Boxplots of observed functional dispersions (FDis) for single functional traits and trait combinations observed at the community scale. 116

Figure 15: Outlying mean index analysis (OMI) of 134 plant species observed in the study area on local environmental variables. 117

Figure 16: Frequencies of significant functional convergence and divergence at the community (plot) and the neighborhood (subplot) scale. 119

Figure 17: Regression trees indicating the relationship between functional convergence and divergence to plot age (age) and soil aeration (SAer). 120

Figure 18: Plots of soil parameters against plot age. 138

Figure 19: Examples of vegetation aspects differing in age and soil aeration. 141

Figure 20: Ordination diagram (principal component analysis) of species of the species pool in the functional space span by traits associated with competitive ability (life span and specific leaf area: SLA) and dispersal ability (seed number/terminal velocity). 147

APPENDIX

Appendix 2: List of tables.

Table 1: Examples for types of assembly rules. 29

Table 2: Decision tree for the metacommunity paradigm expected from the observation or non-observation of relationships between functional traits and dispersal, stress and competition filtering. 32

Table 3: Overview of functional traits considered in this work, and their predicted relevance to describe (i) plant response to dispersal filtering, (ii) plant response to environmental filtering, (iii) plant response to competition filtering, and (iv) the competitive effect of plants exerted on neighbors. 45

Table 4: Spearman correlation coefficients among plant functional traits. 62

Table 5: Results of two-sided Wilcoxon-Mann-Whitney tests for independent samples.. 63

Table 6: Summary of species model averaging results and functional traits............69

Table 7: Summary of soil variables measured at the plot level, with their units, means, standard deviations where applicable, and their eigenvector scores at the 1st axis of the outlying mean index analysis (OMI)............86

Table 8: Results of the Kruskal-Wallis rank sum tests for general differences in establishment success rates among functional groups of three different group definitions. 89

Table 9: Summary of the relationships between the establishment success of functional groups and the availability of vacant space (disturbance and % bare soil), and the covers of functional groups in the resident community. 91

Table 10: Overview of hypothesized functional patterns generated by trait-driven and trait-neutral assembly mechanisms at the community and neighborhood scales. 107

Table 11: Overview of plant traits and trait combinations, and functional importance. 115

Table 12: Summary of soil variables measured at the plot level, with units, means, standard deviations where applicable, and correlation coefficients with the 1st axis of the outlying mean index analysis (OMI). 118

Table 13: Summary of the regression tree analysis results of convergence and divergence occurrences, observed at the community and neighborhood scales, in relationship to plot age (age) and soil aeration (SAer). 122

Table 14: Comparison of relationships between functional traits and dispersal, stress and competition filtering as expected by the four metacommunity paradigms (patch dynamics, neutral, mass effects, species sorting), and confirmed by the results of the present work. 149

Table 15: Summary of predictions of the four metacommunity paradigms (neutral, patch dynamics, species sorting, mass effects). 160

APPENDIX

Appendix 3: List of species in the species pool.

Presented are full species names, life forms sensu Raunkiaer (1934) (P: phanerophyte, C: chamaephyte, H: hemcryptophyte, G: geophyte, T: therophyte, L: liana, V: vascular semi-parasite), seed bank type (t: transient, s: short-term persistent), and further information relevant in this work: Column four shows species categorization to functional groups based on their traits associated with dispersal (seed number/terminal velocity) and competitive ability (life span, specific leaf area): Group A: high competitive ability and high dispersal ability; group B: high competitive ability and low dispersal ability; group C: low competitive ability and intermediate dispersal ability). Columns five to nine indicate that the species was considered as a focal species (occurrence frequency > 10%, see Chapter 4), that it was registered in plots during the vegetation surveys in 2006, 2007, 2008, respectively, and that it was detected in the seed bank samples from 2008. NA denotes missing (trait) information.

Species name	Life form	Seed bank type	Functional group	Focal species	In vegetation 2006	In vegetation 2007	In vegetation 2008	In seed bank 2008
Acer campestre L.	P	t	A			x	x	
Acer platanoides L.	P	t	NA					
Acer pseudoplatanus L.	P	t	A					
Achillea millefolium L.	C	s	A	x	x	x	x	x
Aegopodium podagraria L.	G	t	NA					
Agrostis capillaris L.	H	t	B	x	x	x	x	x
Agrostis gigantea Roth	H	t	B		x			
Agrostis stolonifera L.	H	t	B	x	x	x	x	x
Alchemilla vulgaris L.	H	t	NA					
Allium cepa L.	H	NA	NA					x
Alopecurus geniculatus L.	H	t	C					
Alopecurus pratensis L.	H	NA	B		x	x	x	
Anchusa arvensis (L.) Bieb.	T	t	NA					
Anthoxanthum odoratum L.	H	t	B		x	x	x	
Anthriscus sylvestris (L.) Hoffm.	H	t	B					
Apera spica-venti (L.) Beauv.	T	t	NA			x	x	
Arabidopsis thaliana (L.) Heynh.	H	t	C		x	x	x	
Arabis hirsuta (L.) Scop.	H	t	B		x			
Arctium lappa L.	H	t	A					

167

Appendix 3: Extended.

Species name	Life form	Seed bank type	Functional group	Focal species	In vegetation 2006	2007	2008	In seed bank 2008
Arenaria serpyllifolia L.	C	t	C	x	x	x	x	x
Arrhenatherum elatius (L.) Beauv. ex J. & C. Presl	H	t	B	x	x	x	x	x
Artemisia vulgaris L.	C	s	A	x	x	x	x	x
Bellis perennis L.	H	t	B					
Berteroa incana (L.) DC.	H	t	A		x			
Betula pendula Roth	P	t	NA	x	x	x	x	
Betula pubescens Ehrh.	P	t	NA				x	
Bromus hordeaceus L.	T	t	C		x	x	x	
Bromus inermis Leysser	H	t	B		x	x		
Bromus sterilis L.	T	t	C					
Bromus tectorum L.	T	t	C	x	x	x	x	
Calamagrostis arundinacea (L.) Roth	H	s	A		x			
Calamagrostis epigejos (L.) Roth	H	s	A	x	x	x	x	x
Calystegia sepium (L.) R. Br.	H	t	NA					
Capsella bursa-pastoris (L.) Medicus	T	t	C				x	
Cardamine hirsuta L.	T	t	NA			x	x	x
Carduus crispus L.	H	t	A		x	x	x	
Carex acutiformis Ehrh.	G	t	A					
Carex arenaria L.	H	s	A	x	x	x	x	x
Carex demissa Hornem.	H	t	NA					
Carex flava agg.	H	NA	NA				x	
Carex hirta L.	G	s	B	x	x	x	x	x
Carex ovalis Good.	H	t	NA		x	x		
Carex spicata Hudson	H	t	NA					
Centaurea jacea L.	H	t	A			x		
Cerastium arvense L.	C	t	B					
Cerastium glomeratum Thuill.	T	t	C		x			
Cerastium holosteoides Fr.	H	t	C		x	x	x	x

Appendix 3: Extended.

Species name	Life form	Seed bank type	Functional group	Focal species	In vegetation 2006	In vegetation 2007	In vegetation 2008	In seed bank 2008
Cerastium semidecandrum L.	H	t	C	x	x	x	x	x
Chamomilla suaveolens (Pursh) Rydb.	T	t	NA			x	x	
Chenopodium album L.	T	s	C	x	x	x	x	x
Chenopodium glaucum L.	T	s	C					
Chenopodium polyspermum L.	T	s	NA					
Cirsium arvense (L.) Scop.	G	t	A	x	x	x	x	x
Cirsium oleraceum (L.) Scop.	H	t	NA			x		
Cirsium palustre (L.) Scop.	H	t	A					
Cirsium tuberosum (L.) All.	H	t	NA			x		
Cirsium vulgare (Savi) Ten.	H	t	A	x	x	x	x	x
Clematis vitalba L.	L	t	NA					
Convolvulus arvensis L.	H	t	B					
Conyza canadensis (L.) Cronq.	T	s	C	x	x	x	x	x
Cornus mas L.	P	t	NA			x	x	
Cornus sanguinea L.	P	t	NA					
Corynephorus canescens (L.) Beauv.	H	t	A	x	x	x	x	
Cotoneaster horizontalis Decne	P	NA	NA		x	x	x	
Crataegus monogyna ssp. *monogyna*	P	NA	NA				x	
Crepis biennis L.	H	t	C					
Crepis capillaris (L.) Wallr.	H	t	C		x	x	x	
Cytisus scoparius (L.) Link	P	t	NA					
Dactylis glomerata L.	H	t	B	x	x	x	x	
Daucus carota L.	H	t	A	x	x	x	x	x
Deschampsia cespitosa (L.) Beauv.	H	t	A	x	x	x	x	
Deschampsia flexuosa (L.)	H	t	A		x		x	

Appendix 3: Extended.

Species name	Life form	Seed bank type	Functional group	Focal species	In vegetation 2006	2007	2008	In seed bank 2008
Trin.								
Echium vulgare L.	H	t	A					
Elymus pungens (Pers.) Melderis	G	t	B	x	x	x	x	x
Epilobium angustifolium L.	H	t	A		x	x	x	x
Epilobium ciliatum Rafin.	H	t	C					
Epilobium hirsutum L.	H	s	A		x			x
Epilobium palustre L.	H	t	C					
Epilobium parviflorum Schreber	H	NA	NA		x		x	
Epilobium tetragonum L.	H	t	A					
Equisetum arvense L.	G	NA	NA		x	x	x	x
Equisetum palustre L.	G	NA	NA					
Erigeron acer L.	H	t	A					
Erodium cicutarium (L.) L'Hér.	H	t	C		x	x	x	
Erophila verna (L.) Chevall.	T	t	C		x	x	x	x
Euphorbia cyparissias L.	H	t	B					
Festuca arundinacea Schreber	H	t	B		x	x		
Festuca ovina L.	H	t	C	x	x	x	x	
Festuca pratensis Hudson	H	t	B			x	x	
Festuca rubra L.	H	t	C	x	x	x	x	
Fragaria vesca L.	H	t	NA					
Fraxinus excelsior L.	P	t	A			x	x	
Galeopsis tetrahit L.	T	t	C					
Galium aparine L.	L	t	C				x	
Galium mollugo L.	H	t	B			x		
Geranium molle L.	T	t	NA		x	x	x	
Geranium robertianum L.	H	t	C					
Geranium rotundifolium L.	T	t	NA					
Geum urbanum L.	H	t	B					
Glechoma hederacea L.	H	t	NA		x	x	x	

170

APPENDIX

Appendix 3: Extended.

Species name	Life form	Seed bank type	Functional group	Focal species	In vegetation 2006	2007	2008	In seed bank 2008
Gnaphalium uliginosum L.	T	NA	NA					x
Heracleum sphondylium L.	H	t	A					
Herniaria glabra L.	H	NA	A			x		
Hieracium laevigatum Willd.	H	t	A					
Hieracium pilosella L.	H	t	A		x	x	x	
Hieracium praealtum ssp. *bauhinii* (Besser) Petunnikov	H	t	NA					
Hippophae rhamnoides L.	P	t	NA		x		x	
Holcus lanatus L.	H	s	B	x	x	x	x	x
Holcus mollis L.	H	t	B		x			
Hypericum perforatum L.	H	s	B	x	x	x	x	x
Hypochoeris radicata L.	H	t	A	x	x	x	x	x
Jasione montana L.	H	t	C					
Juncus articulatus L.	H	t	A			x		
Juncus bufonius L.	T	s	NA				x	x
Juncus compressus Jacq.	G	t	NA		x			
Juncus conglomeratus L.	H	t	A					
Juncus effusus L.	H	s	A		x	x		
Juncus tenageia L. fil.	T	t	NA		x			
Juncus tenuis Willd.	H	t	NA		x			
Lactuca serriola L.	H	t	NA					
Lamium album L.	H	t	B					
Lamium purpureum L.	H	t	NA					
Lathyrus pratensis L.	H	t	B		x	x	x	
Leontodon autumnalis L.	H	t	B		x	x		
Lepidium campestre (L.) R. Br.	T	t	C					
Lepidium densiflorum Schrader	H	NA	NA					
Leucanthemum vulgare Lam.	H	t	B				x	
Logfia minima (Sm.)	T	NA	C		x	x	x	x

Appendix 3: Extended.

Species name	Life form	Seed bank type	Functional group	Focal species	In vegetation 2006	In vegetation 2007	In vegetation 2008	In seed bank 2008
Dumort.								
Lolium perenne L.	H	t	B	x	x	x	x	
Lotus corniculatus L.	H	t	B	x	x	x	x	x
Lotus uliginosus Schkuhr	H	t	NA					
Luzula campestris (L.) DC.	H	t	B			x	x	
Luzula luzuloides (Lam.) Dandy & Wilmott	H	t	B					
Luzula sylvatica (Hudson) Gaudin	H	t	NA				x	
Lythrum salicaria L.	H	t	NA					
Mahonia aquifolium (Pursh) Nutt.	C	NA	NA					
Malva neglecta Wallr.	H	t	NA					
Matricaria maritima L.	T	t	A	x	x	x	x	x
Matricaria recutita L.	T	NA	NA					x
Medicago arabica (L.) Hudson	T	t	NA		x			
Medicago lupulina L.	H	s	B	x	x	x	x	x
Medicago sativa L.	H	t	B					
Melilotus alba Medicus	H	t	NA		x			
Melilotus officinalis (L.) Pallas	H	t	C	x	x	x	x	x
Myosotis arvensis (L.) Hill	H	t	C		x	x		x
Myosotis ramosissima Rochel	T	t	NA		x	x	x	
Myosotis scorpioides L.	H	t	NA		x			
Myosotis stricta Link ex Roemer & Schultes	T	t	C					
Oenothera biennis L.	H	s	A	x	x	x	x	x
Ornithopus perpusillus L.	T	t	NA			x	x	
Papaver dubium L.	T	t	NA		x			
Papaver somniferum L.	T	t	NA					
Phalaris arundinacea L.	H	t	B	x	x	x	x	
Phleum pratense L.	H	t	NA		x	x	x	
Phragmites australis (Cav.)	G	t	A		x	x	x	

Appendix 3: Extended.

Species name	Life form	Seed bank type	Functional group	Focal species	In vegetation 2006	2007	2008	In seed bank 2008
Trin. ex Steudel								
Picris hieracioides L.	H	t	C		x			
Pinus nigra Arnold	P	t	NA					
Plantago lanceolata L.	H	t	A	x	x	x	x	x
Plantago major L.	H	s	A	x		x	x	x
Plantago media L.	H	s	A		x			
Poa angustifolia L.	H	t	A					
Poa annua L.	H	s	C		x	x	x	x
Poa compressa L.	H	t	A			x	x	
Poa nemoralis L.	H	t	B			x	x	
Poa palustris L.	H	t	B		x			
Poa pratensis L.	H	s	B	x	x	x	x	x
Poa trivialis L.	H	t	C		x	x	x	x
Polygonum amphibium L.	G	s	NA					
Polygonum aviculare L.	T	NA	C		x	x		x
Polygonum lapathifolium L.	T	s	NA					
Polygonum persicaria L.	T	s	C	x	x	x	x	x
Populus alba L.	P	NA	NA		x	x	x	
Populus tremula L.	P	t	NA			x	x	
Potentilla anserina L.	H	t	B		x	x	x	x
Potentilla inclinata Vill.	H	NA	NA					
Potentilla intermedia L.	H	t	NA		x			
Potentilla norvegica L.	H	s	NA					
Potentilla recta L. sensu stricto	H	t	A					
Potentilla reptans L.	H	t	B		x	x	x	
Prunella vulgaris L.	H	t	B		x		x	
Prunus padus L.	P	t	NA					
Prunus serotina Ehrh.	P	t	NA					
Quercus robur L.	P	t	NA		x	x	x	
Ranunculus acris L.	H	t	B			x	x	
Ranunculus arvensis L.	T	t	C					

Appendix 3: Extended.

Species name	Life form	Seed bank type	Functional group	Focal species	In vegetation 2006	2007	2008	In seed bank 2008
Ranunculus repens L.	H	t	B		x			x
Ranunculus sceleratus L.	T	s	NA					x
Reseda lutea L.	H	t	A			x		
Rhinanthus angustifolius C.C. Gmelin	V	t	C		x	x	x	
Ribes rubrum L.	P	t	NA					
Robinia pseudacacia L.	P	t	NA					
Rorippa palustris (L.) Besser	H	t	C					
Rosa gallica L.	P	NA	NA					
Rubus caesius L.	C	t	NA				x	
Rubus fruticosus agg. L.	P	t	NA		x			
Rubus idaeus L.	P	t	B					
Rumex acetosella L.	G	s	B	x	x		x	x
Rumex crispus L.	H	t	A		x	x	x	
Rumex obtusifolius L.	H	t	C			x	x	
Sagina procumbens L.	C	s	A		x	x	x	x
Salix acutifolia Willd.	P	NA	NA		x	x		
Salix alba L.	P	t	NA			x	x	
Salix aurita L.	P	t	NA			x	x	
Salix caprea L.	P	t	A		x	x	x	
Salix cinerea L.	P	t	NA					
Salix elaeagnos Scop.	P	NA	NA			x		
Salix glabra Scop.	C	NA	NA			x		
Salix purpurea L.	P	t	NA		x	x		
Salix viminalis L.	P	t	A		x	x	x	
Sanguisorba minor Scop.	H	t	B		x	x	x	
Saponaria officinalis L.	H	t	B					
Saxifraga tridactylites L.	T	t	C			x	x	
Scleranthus annuus L.	T	t	NA			x		
Sedum acre L.	C	t	NA			x	x	
Senecio inaequidens DC.	C	t	A	x	x	x	x	x
Senecio viscosus L.	T	t	C					

APPENDIX

Appendix 3: Extended.

Species name	Life form	Seed bank type	Functional group	Focal species	In vegetation 2006	In vegetation 2007	In vegetation 2008	In seed bank 2008
Senecio vulgaris L.	H	t	C		x	x	x	
Setaria viridis (L.) Beauv.	T	NA	NA					x
Silene latifolia Poiret	H	t	NA					
Sisymbrium altissimum L.	H	s	C	x	x			
Sisymbrium officinale (L.) Scop.	T	t	NA					
Solanum dulcamara L.	L	t	NA					
Solanum nigrum L.	T	NA	NA					x
Solidago canadensis L.	G	t	A					x
Solidago gigantea Aiton	G	t	A	x	x	x	x	x
Solidago virgaurea L.	H	t	A					
Sonchus asper (L.) Hill	T	t	C		x	x		x
Sonchus oleraceus L.	T	t	NA			x	x	
Spergula morisonii Boreau	T	t	A					
Spergularia media (L.) C. Presl	C	t	NA		x	x		
Spergularia rubra (L.) J. & C. Presl	H	s	C					
Stachys germanica L.	H	NA	NA					
Stellaria graminea L.	H	t	B					
Stellaria media (L.) Vill.	C	t	C			x		x
Symphytum officinale L.	H	t	NA					
Tanacetum vulgare L.	H	s	A	x	x	x	x	x
Taraxacum officinale agg.	H	t	B	x	x	x	x	x
Thlaspi arvense L.	T	t	C					
Tragopogon dubius Scop.	H	t	C			x	x	
Trifolium arvense L.	T	t	C	x	x	x	x	x
Trifolium campestre Schreber	T	t	C	x	x	x	x	x
Trifolium fragiferum L.	H	t	B			x	x	
Trifolium pratense L.	H	t	B	x	x	x	x	x
Trifolium repens L.	C	s	B	x	x	x	x	x
Trifolium scabrum L.	T	NA	NA					
Tussilago farfara L.	G	t	A	x	x	x	x	

Appendix 3: Extended.

Species name	Life form	Seed bank type	Functional group	Focal species	In vegetation 2006	2007	2008	In seed bank 2008
Urtica dioica L.	H	t	NA					
Verbascum densiflorum Bertol.	H	t	NA					
Verbascum thapsus L.	H	t	NA					
Veronica arvensis L.	T	t	C		x	x	x	x
Veronica serpyllifolia L.	H	t	NA		x		x	
Veronica verna L.	T	t	NA					
Vicia cracca L.	H	t	B		x	x		
Vicia hirsuta (L.) S.F. Gray	L	t	C	x	x	x	x	x
Vicia sativa ssp. *nigra* (L.) Ehrh.	L	t	C	x	x	x	x	x
Vicia tenuissima (Bieb.) Schinz & Thell.	L	NA	NA					
Vicia tetrasperma (L.) Schreber	L	t	C		x	x	x	
Vicia villosa Roth	H	t	C		x			
Vulpia myuros (L.) C.C. Gmelin	T	s	C	x	x	x	x	x

9 References

Ackerly, D. D. and P. B. Reich (1999). Convergence and correlations among leaf size and function in seed plants: A comparative test using independent contrasts. Am. J. Bot. 86(9): 1272-1281.

Ad-hoc-AG Boden (2005). Bodenkundliche Kartieranleitung. Stuttgart, Schweizerbart.

Adler, P. B., J. HilleRisLambers and J. M. Levine (2007). A niche for neutrality. Ecol. Lett. 10(2): 95-104.

Aerts, R. (1999). Interspecific competition in natural plant communities: Mechanisms, trade-offs and plant-soil feedbacks. J. Exp. Bot. 50: 29-37.

Amarasekare, P. (2003). Competitive coexistence in spatially structured environments: A synthesis. Ecol. Lett. 6: 1109-1122.

Amarasekare, P., M. F. Hoopes, N. Mouquet and M. Holyoak (2004). Mechanisms of coexistence in competitive metacommunities. Am. Nat. 164(3): 310-326.

Angold, P. G., J. P. Sadler, M. O. Hill, A. Pullin, S. Rushton, K. Austin, E. Small, B. Wood, R. Wadsworth, R. Sanderson and K. Thompson (2006). Biodiversity in urban habitat patches. Sci. Total Environ. 360(1-3): 196-204.

Anten, N. P. R. and T. Hirose (1999). Interspecific differences in above-ground growth patterns result in spatial and temporal partitioning of light among species in a tall-grass meadow. J. Ecol. 87(4): 583-597.

Armbruster, W. S., M. E. Edwards and E. M. Debevec (1994). Floral character displacement generates assemblage structure of Western-Australian triggerplants (Stylidium). Ecology 75(2): 315-329.

Baer, S. G., J. M. Blair, S. L. Collins and A. K. Knapp (2004). Plant community responses to resource availability and heterogeneity during restoration. Oecologia 139(4): 617-629.

Bakker, J. P. (1989). Nature management by grazing and cutting: On the ecological significance of grazing and cutting regimes applied to restore former species-rich grassland communities in the Netherlands. Dordrecht Kluwer Acad. Publ.

Banta, J. A., S. C. Stark, M. H. H. Stevens, T. H. Pendergast, A. Baumert and W. P. Carson (2008). Light reduction predicts widespread patterns of dominance between asters and goldenrods. Plant Ecol. 199(1): 65-76.

Bastin, L. and C. D. Thomas (1999). The distribution of plant species in urban vegetation fragments. Landscape Ecol. 14: 493-507.

Bekker, R. M., J. P. Bakker, U. Grandin, R. Kalamees, P. Milberg, P. Poschlod, K. Thompson and J. H. Willems (1998). Seed size, shape and vertical distribution in the soil: Indicators of seed longevity. Funct. Ecol. 12: 834-842.

Bell, G. (2001). Neutral Macroecology. Science 293: 2413-2418.

Bell, G. (2005). The co-distribution of species in relation to the neutral theory of community ecology. Ecology 86(7): 1757-1770.

Borcard, D., P. Legendre and P. Drapeau (1992). Partialling out the spatial component of ecological variation. Ecology 73(3): 1045-1055.

Bossuyt, B. and O. Honnay (2006). Interactions between plant life span, seed dispersal capacity and fecundity determine metapopulation viability in a dynamic landscape. Landscape Ecol. 21: 1195-1205.

Breiman, L., J. H. Friedmann, R. A. Olshen and C. G. Stone (1984). Classification and regression trees. Belmont, California, USA, Wadsworth International Group.

Brooker, R. W. (2006). Plant-plant interactions and environmental change. New Phytol.(171): 271-284.

Brown, J. H. (1973). Species diversity of seed-eating desert rodents in sand dune habitats. Ecology 54: 775-787.

Brown, J. H. and A. Kodric-Brown (1977). Turnover rates in insular biogeography: Effect of immigration on extinction. Ecol. Soc. 58(2): 445-449.

Brown, J. H. and A. Kodric-Brown (1979). Convergence, competition, and mimicry in a temperate community of hummingbird-pollinated flowers. Ecology 60(5): 1022-1035.

Brown, W. L. and E. O. Wilson (1956). Character displacement. Systematic Zoology 5: 49-65.

Bruno, J. F., J. J. Stachowicz and M. D. Bertness (2003). Inclusion of facilitation into ecological theory. Trends Ecol. Evol. 18(3): 119-125.

Bullock, J. M. (2000). Gaps and seedling colonization. In: M. Fenner. Seeds. The ecology of regeneration in plant communities. Wallingford, UK, CABI Publishing: 375-395.

Burke, M. J. W. and J. P. Grime (1996). An experimental study of plant community invasibility. Ecology 77(3): 776-790.

Burnham, K. P. and D. R. Anderson (2002). Model selection and multimodel inference: a practical information-theoretic approach. New York, Springer.

Burns, K. C. (2007). Patterns in the assembly of an island plant community. J. Biogeogr. 34: 760–768.

Burns, K. C. and C. J. Neufeld (2009). Plant extinction dynamics in an insular metacommunity. Oikos 118 (2): 191-198.

Byer, M. D. (1969). The role of physical environment in some tracheophyte distributions along a soil moisture gradient. Bulletin of the Torrey Botanical Club 96(2): 191-201.

Cadotte, M. W., J. Cavender-Bares, D. Tilman and T. H. Oakley (2009). Using phylogenetic, functional and trait diversity to understand patterns of plant community productivity. PLoS ONE 4(5): e5695.

Calcagno, V., N. Mouquet, P. Jarne and P. David (2006). Coexistence in a metacommunity: The competition-colonization trade-off is not dead. Ecol. Lett. 9(8): 897-907.

Calinski, T. and J. Harabasz (1974). A dendrite method for cluster analysis. Commun. Stat. 3: 1-27.

Caswell, H. (1976). Community structure: A neutral model analysis. Ecol. Monogr. 46(3): 327-354.

Chase, J. M. (2007). Drought mediates the importance of stochastic community assembly. PNAS 104(44): 17430-17434.

Chase, J. M., P. Amarasekare, K. Cottenie, A. Gonzalez, R. D. Holt, M. Holyoak, M. F. Hoopes, M. A. Leibold, M. Loreau, N. Mouquet, J. B. Shurin and D. Tilman (2005). Competing theories for competitive metacommunities. In: M. Holyoak, M. A. Leibold and R. D. Holt. Metacommunities. Spatial dynamics and ecological communities. Chicago, The University of Chicago: 335-354.

Chave, J. and E. G. Leigh (2002). A spatially explicit neutral model of beta-diversity in tropical forests. Theor. Pop. Biol. 62: 153-168.

Chave, J., H. Muller-Landau and S. A. Levin (2002). Comparing classical community modules: Theoretical consequences for patterns of diversity. Am. Nat. 159: 1-23.

Cingolani, Cabido, Gurvich, Renison and Díaz (2007). Filtering processes in the assembly of plant communities: Are species presence and abundance driven by the same traits? J. Veg. Sc. 18: 911-920.

Clark, J. S. (2008). Beyond neutral science. Trends Ecol. Evol. 24 (1): 8-15.

Clarke, P. J. and E. A. Davison (2004). Emergence and survival of herbaceous seedlings in temperate grassy woodlands: Recruitment limitations and regeneration niche. Austral Ecol. 29: 320-331.

Clements, F. E. (1916). Plant succession. Washington, DC, Carnegie Institute of Washington.

Connell, J. H. and R. O. Slatyer (1977). Mechanisms of succession in natural communities and their role in community stability and organization. Am. Nat. 111(982): 1119-1144.

Cook, W. M., J. Yao, B. L. Foster, R. D. Holt and L. B. Patrick (2005). Secondary succession in an experimentally fragmented landscape: Community patterns across space and time. Ecology 86(5): 1267-1279.

Cornwell, W. K. and D. D. Ackerly (2009). Community assembly and shifts in plant trait distributions across an environmental gradient in coastal California. Ecol. Monogr. 79(1): 109-126.

Cornwell, W. K., D. W. Schwilk and D. D. Ackerly (2006). A trait-based test for habitat filtering: Convex hull volume. Ecology 87(6): 1465-1471.

Cottenie, K. (2005). Integrating environmental and spatial processes in ecological community dynamics. Ecol. Lett. 8: 1175-1182.

Cottenie, K. and L. De Meester (2005). Local interactions and local dispersal in a Zooplankton metacommunity. In: M. Holyoak, M. A. Leibold and R. D. Holt.

Metacommunities. Spatial dynamics and ecological communities. Chicago, The University of Chicago: 189-211.

Cottenie, K., E. Michels, N. Nuytten and L. De Meester (2003). Zooplankton metacommunity structure: Regional vs. local processes in highly interconnected ponds. Ecology 84: 991-1000.

Crawley, M. J. (2005). Statistics. An introduction using R. Chichester, John Wiley & Sons Ltd.

Crawley, M. J., P. H. Harvey and A. Purvis (1996). Comparative ecology of the native and alien floras of the British Isles. Philosophical Transactions: Biological Sciences 351(1345): 1251-1259.

Cunningham, S. A., B. Summerhayes and M. Westoby (1999). Evolutionary divergences in leaf structure and chemistry, comparing rainfall and soil nutrient gradients. Ecol. Monogr. 69(4): 569-588.

Darwin, C. (1859). The origin of species by means of natural selection, or the preservation of favoured races in the struggle for life. London, John Murray.

Davies, A. and S. Waite (1998). The persistence of calcareous grassland species in the soil seed bank under developing and established scrub. Plant Ecol. 136(1): 27-39.

Davis, M., J. P. Grime and K. Thompson (2000). Fluctuating resources in plant communities: A general theory of invasibility. J. Ecology 88: 528-534.

Dayan, T. and D. Simberloff (1994). Morphological relationships among coexisting heteromyids: An incisive dental character. Am. Nat. 143: 462-477.

de Bello, F., J. Lepš and M.-T. Sebastià (2005). Predictive value of plant traits to grazing along a climatic gradient in the Mediterranean. J. Appl. Ecol. 42.

de Bello, F., W. Thuiller, J. Lepš, P. Choler, J. C. Clément, P. Macek, M. T. Sebastiá and S. Lavorel (2009). Partitioning of functional diversity reveals the scale and extent of trait convergence and divergence. J. Veg. Sc 20(3): 475 - 486.

de Meester, L., G. Louette, C. Duvivier, C. Damme and E. Michels (2007). Genetic composition of resident populations influences establishment success of immigrant species. Oecol. Aquat. 153(2): 431-440.

del Moral, R. and S. Y. Grishin (1999). Volcanic disturbances and ecosystem recovery. In: L. R. Walker. Ecosystems of disturbed ground. New York, USA, Elsevier: 137-160.

Deutscher Wetterdienst. (2006/07). Mittelwerte des Niederschlags und der Temperatur für den Zeitraum 1961-1990. downloaded April 06 2010, from http://www.dwd.de/bvbw.

Diamond, J. M. (1975). Assembly of species communities. In: J. M. Diamond and M. L. Cody. Ecology and evolution of communities. Cambridge, MA, Harvard University Press: 342-444.

Díaz, S. and M. Cabido (1997). Plant functional types and ecosystem function in relation to global change. J. Veg. Sci. 8(4): 463-474.

Díaz, S., M. Cabido and F. Casanoves (1999). Functional implications of trait-environment linkages in plant communities. In: E. Weiher and P. Keddy. Ecological assembly rules: Perspectives, advances, retreats. Cambridge, UK, Cambridge University Press: 338-392.

Díaz, S., J. G. Hodgson, K. Thompson, M. Cabido, J. H. C. Cornelissen, A. Jalili, G. Montserrat-Martí, J. P. Grime, F. Zarrinkamar, Y. Asri, S. R. Band, S. Basconcelo, P. Castro-Díez, G. Funes, B. Hamzehee, M. Khoshnevi, N. Pérez-Harguindeguy, M. C. Pérez-Rontomé, F. A. Shirvany, F. Vendramini, S. Yazdani, R. Abbas-Azimi, A. Bogaard, S. Boustani, M. Charles, M. Dehghan, L. de Torres-Espuny, V. Falczuk, J. Guerrero-Campo, A. Hynd, G. Jones, E. Kowsary, F. Kazemi-Saeed, M. Maestro-Martínez, A. Romo-Díez, S. Shaw, B. Siavash, P. Villar-Salvador and M. R. Zak (2004). The plant traits that drive ecosystems: Evidence from three continents. J. Veg. Sci. 15(3): 295-304.

Dirnböck, T. and S. Dullinger (2004). Habitat distribution models, spatial autocorrelation, functional traits and dispersal capacity of alpine plant species. J. Veg. Sci. 15(1): 77-84.

Dobzhansky, T. G. (1951). Genetics and the origin of species. Columbia University Press, New York.

Dolédec, S., D. Chessel and C. Gimaret-Carpentier (2000). Niche separation in community analysis: A new method. Ecology 81(10): 2914-2927.

Dormann, C., J. McPherson, M. Araújo, R. Bivand, J. Boilliger, G. Carl, R. G. Davies, A. Hirzel, W. Jetz, W. Kissling, I. Kühn, R. Ohlemüller, P. Peres-Neto, B. Reineking, B. Schröder, F. Schurr and R. Wilson (2007). Methods to account for spatial autocorrelation in the analysis of species distributional data: A review. Ecography 30: 609-628.

Dray, S. (2006). Moran's eigenvectors of spatial weighting matrices in R. SpacemakeR Tutorial, http://biomserv.univ-lyon1.fr/~dray/software.php.

Dray, S., P. Legendre and P. R. Peres-Neto (2006). Spatial modelling: A comprehensive framework for principal coordinate analysis of neighbour matrices (PCNM). Ecol. Model. 196: 483–493.

Driscoll, D. A. and D. B. Lindenmayer (2009). Empirical tests of metacommunity theory using an isolation gradient. Ecol. Monogr. 79(3): 485-501.

Duckworth, J. C., M. Kent and P. M. Ramsay (2000). Plant functional types: An alternative to taxonomic plant community description in biogeography? Progress in Physical Geography 24(4): 515-542.

Dukes, J. S. (2001). Biodiversity and invasibility in grassland microcosms. Oecologia 126(4): 563-568.

Dupré, C. and J. Ehrlén (2002). Habitat configuration, species traits and plant distributions. J. Ecol. 90(5): 796-805.

Egnér, H., H. Riehm and W. R. Domingo (1960). Untersuchungen über die Bodenanalyse als Grundlage für die Beurteilung des Nährstoffzustandes des Bodens II. Chemische Extraktionsmethoden zur Phosphor- und Kaliumbestimmung. Kungl. Lantbrukshögskolans Annualer 26: 199-215.

Ehrlén, J. and O. Eriksson (2003). Large-scale spatial dynamics of plants: A response to Freckleton & Watkinson. J. Ecology 91: 316-320.

Ehrlén, J., Z. Münzbergová, M. Diekmann and O. Eriksson (2006). Long-term assessment of seed limitation in plants: results from an 11-year experiment. J. Ecol. 94(6): 1224-1232.

Ehrlén, J. and J. M. van Groenendael (1998). The trade-off between dispersability and longevity - an important aspect of plant species diversity. Appl. Veg. Sci. 1: 29-36.

Ejrnaes, R., H. H. Bruun and B. J. Graae (2006). Community assembly in experimental grasslands: suitable environment or timely arrival? Ecology 87(5): 1225-1233.

Ellenberg, H. (1986). Vegetation Mitteleuropas mit den Alpen in ökologischer Sicht. Stuttgart, Ulmer.

Ellis, A. M., L. P. Lounibos and M. Holyoak (2006). Evaluating the long-term metacommunity dynamics of tree hole mosquitoes. Ecology 87(10): 2582-2590.

Elton, C. S. (1958). The ecology of invasions by animals and plants. London, Chapman and Hall.

Elton, C. S. (1972). Animal Ecology, Chapman and Hall Ltd.

Emery (2007). Limiting similarity between invaders and dominant species in herbaceous plant communities? J. Ecology 95: 1027-1035.

Endels, P., D. Adriaens, R. M. Bekker, I. C. Knevel, G. Decocq and M. Hermy (2007). Groupings of life-history traits are associated with distribution of forest plant species in a fragmented landscape. J. Veg. Sc 18: 499-508.

Enquist, B. J., J. Sanderson and M. D. Weiser (2002). Modeling macroscopic patterns in ecology. Science 295: 1835-1836.

Eriksson, O. (1996). Regional dynamics of plants: A review of evidence for remnant, source-sink and metapopulations. Oikos 77(2): 183-237.

Eschtruth, A. K. and J. J. Battles (2009). Assessing the relative importance of disturbance, herbivory, diversity, and propagule pressure in exotic plant invasion. Ecol. Monogr. 79(2): 265-280.

ESRI Inc. (2006). ArcMap^{TM}. Redlands, California.

Ewers, R. and R. Didham (2006). Confounding factors in the detection of species responses to habitat fragmentation. Biol. Rev. 81(1): 117-142.

Fahrig, L. (2003). Effects of habitat fragmentation on biodiversity. Annual review of ecology, evolution, and systematics 34: 487-515.

Fargione, J., C. S. Brown and D. Tilman (2003). Community assembly and invasion: An experimental test of neutral versus niche processes. Proceedings of the National Academy of Sciences 100(15): 8916–8920.

Fenner, M. (1978). Susceptibility to shade in seedlings of colonizing and closed turf species. New Phytol. 81(3): 739-744.

Fenner, M. (1987). Seed characteristics in relation to succession. In: A. J. Gray, M. J. Crawley and P. J. Edwards. Colonization, succession and stability. Oxford London Edinburgh Boston Palo Alto Melbourne, Blackwell Scientific Publications: 103-114.

Fenner, M. and K. Thompson (2004). The ecology of seeds. Cambridge, Cambridge University Press.

Fonseca, C. R., J. M. Overton, B. Collins and M. Westoby (2000). Shifts in trait-combinations along rainfall and phosphorus gradients. J. Ecol. 88(6): 964-977.

Foster, B. L., T. L. Dickson, C. A. Murphy, I. S. Karel and V. H. Smith (2004). Propagule pools mediate community assembly and diversity-ecosystem regulation along a grassland productivity gradient. J. Ecol. 92(3): 435 - 449.

Fox, B. J. (1987). Species assembly and the evolution of community structure. Evol. Ecol. 1: 201-213.

Freckleton, R. P. and A. R. Watkinson (2002). Large-scale spatial dynamics of plants: Metapopulations, regional ensembles and patchy populations. J. Ecol. 90: 419-434.

Freckleton, R. P. and A. R. Watkinson (2003). Are all plant populations metapopulations? J. Ecology 91: 321-324.

Freestone, A. L. and B. D. Inouye (2006). Dispersal limitation and environmental heterogeneity shape scale-dependent diversity patterns in plant communities. Ecology 87(10): 2425-2432.

Frelich, L. E. and P. B. Reich (1999). Neighborhood effects, disturbance severity, and community stability in forests. Ecosystems 2: 151-166.

Fridley, J. D., R. L. Brown and J. F. Bruno (2004). Null models of exotic invasion and scale-dependent patterns of native and exotic species richness. Ecology 85(12): 3215–3222.

Garnier, E. (1992). Growth analysis of congeneric annual and perennial grass species. J. Ecol. 80(4): 665-675.

Garnier, E., P. Cordonnier, J.-L. Guillerm and L. Sonié (1997). Specific leaf area and leaf nitrogen concentration in annual and perennial grass species growing in Mediterranean old-fields. Oecologia 111: 490-498.

Gaudet, C. L. and P. A. Keddy (1988). A comparative approach to predicting competitive ability from plant traits. Nature 334: 242 - 243.

Gewin, V. (2006). Beyond neutrality - ecology finds its niche. PLoS Biol.

Gilbert, B. and M. J. Lechowicz (2004). Neutrality, niches, and dispersal in a temperate forest understory. PNAS 101 (20): 7651-7656.

Girdler, E. B. and B. T. C. Barrie (2008). The scale-dependent importance of habitat factors and dispersal limitation in structuring Great Lakes shoreline plant communities. Plant Ecol. 198: 211-223.

Gitay, H. and I. R. Noble (1997). What are functional types and how should we seek them? In: T. M. Smith, H. H. Shugart and F. I. Woodward. Plant functional types: Their relevance to ecosystem properties and global change. Cambridge, Cambridge University Press: 3-19.

Gitay, H. and J. B. Wilson (1995). Postfire changes in community structure of tall tussock grasslands - a test of alternative models of succession. J. Ecol. 83(5): 775-782.

Godefroid, S. and N. Koedam (2007). Urban plant species patterns are highly driven by density and function of built-up areas. Landscape Ecol. 22(8): 1227-1239.

Godefroid, S., D. Monbaliu and N. Koedam (2007). The role of soil and microclimatic variables in the distribution patterns of urban wasteland flora in Brussels, Belgium. Landscape Urban Plann. 80(1-2): 45-55.

Goldberg, D. E. (1996). Competitive ability: Definitions, contingency and correlated traits. Phil. Trans. R. Soc. B. 351: 1377-1385.

Goldberg, D. E. and A. M. Barton (1992). Patterns and consequences of interspecific competition in natural communities - a review of field experiments with plants. Am. Nat. 139: 771-801.

Goldberg, D. E. and P. A. Werner (1983). Equivalence of competitors in plant communities: A null hypothesis and a field experimental approach. Am. J. Bot. 70(7): 1098-1104.

Gonzalez, A. (2005). Local and regional community dynamics in fragmented landscapes. Insights from a bryophyte-based natural microcosm. In: M. Holyoak, M. A. Leibold and R. D. Holt. Metacommunities. Spatial dynmics and ecological communities. Chicago, The University of Chicago: 146-169.

Gotelli, N. J. and D. J. McCabe (2002). Species co-occurrence: A meta-analysis of J. M. Diamond's assembly rules model. Ecology 83(8): 2091-2096.

Gotelli, N. J. and B. J. McGill (2006). Null versus neutral models: What's the difference? Ecography 29(5): 793-800.

Grashof-Bokdam, C. J. and W. Geertsema (1998). The effect of isolation and history on colonization patterns of plant species in secondary woodland. J. Biogeogr. 25(5): 837-846.

Griffith, D. A. and P. R. Peres-Neto (2006). Spatial modeling in ecology: The flexibility of eigenfunction spatial analyses. Ecology 87(10): 2603-2613.

Grime, J. P. (1965). Shade tolerance in flowering plants. Nature 208: 161.

Grime, J. P. (2002). Plant strategies, vegetation processes, and ecosystem properties. Chichester, John Wiley & Sons.

Grime, J. P. (2006). Trait convergence and trait divergence in herbaceous plant communities: Mechanisms and consequences. J. Vegetation Science 17: 255-260.

Grime, J. P., J. G. Hodgson and R. Hunt (1989). Comparative plant ecology: A functional approach to common British species. London, Unwin Hyman.

Gross, K. L. and P. A. Werner (1982). Colonizing abilities of "biennial" plant species in relation to ground cover: Implications for their distributions in a successional sere. Ecology 63(4): 921-931.

Gross, N., K. N. Suding and S. Lavorel (2007). Leaf dry matter content and lateral spread predict response to land use change for six subalpine grassland species. J. Veg. Sci. 18(2): 289-300.

Grubb, P. J. (1977). The maintenace of species-richness in plant communities: The importance of the regeneration niche. Biol. Rev. 52(1): 107-145.

Grubb, P. J. (1987). Some generalizing ideas about colonization and succession in green plants and fungi. In: A. J. Gray, M. J. Crawley and P. J. Edwards. Colonization, succession and stabiilty. Oxford London Edinburgh Boston Palo Alto Melbourne, Blackwell Scientific Publications: 81-102.

Halpern, C. B., J. A. Antos, M. A. Geyer and A. M. Olson (1997). Species replacement during early secondary succession: The abrupt decline of a winter annual. Ecology 78(2): 621-631.

Hanski, I. (1994). A practical model of metapopulation dynamics. J. Anim. Ecol. 63(1): 151-162.

Hanski, I. and M. Gilpin (1991). Metapopulation dynamics: Brief history and conceptual domain. Biol. J. Linn. Soc. 42(1-2): 3-16.

Harpole, W. S. and D. Tilman (2006). Non-neutral patterns of species abundance in grassland communities. Ecol. Lett. 9: 15–23.

Hastings, A. (1980). Disturbance, coexistence, history, and competition for space. Theor. Popul. Biol. 18: 363-373.

Helm, A., I. Hanski and M. Pärtel (2006). Slow response of plant species richness to habitat loss and fragmentation. Ecol. Lett. 9(1): 72-77.

Henle, K., K. F. Davies, M. Kleyer, C. Margules and J. Settele (2004). Predictors of species sensitivity to fragmentation. Biodivers. Conserv. 13: 207-251.

Hérault, B. and O. Honnay (2005). The relative importance of local, regional and historical factors determining the distribution of plants in fragmented riverine forests: An emergent group approach. J. Biogeogr. 32(12): 2069-2081.

Higgins, S. I., R. Nathan and M. L. Cain (2003). Are long-distance dispersal events in plants usually caused by nonstandard means of dispersal? Ecology 84(8): 1945–1956.

Hodgson, J. G., P. J. Wilson, R. Hunt, J. P. Grime and K. Thompson (1999). Allocating C-S-R plant functional types: A soft approach to a hard problem. Oikos 85: 282-294.

Holdaway, R. J. and A. D. Sparrow (2006). Assembly rules operating along a primary riverbed-grassland successional sequence. J. Ecol. 94(6): 1092-1102.

Holt, R. D. (2001). Species coexistence. Encyclopedia of Biodiversity 5: 413–426.

Holyoak, M., M. A. Leibold and R. D. Holt (2005). Metacommunities. Spatial dynamics and ecological communities. Chicago, The University of Chicago.

Horn, H. S. and R. H. MacArthur (1972). Competition among fugitive species in a harlequin environment. Ecol. Soc. 53: 749-752.

Howard, T. G. and D. E. Goldberg (2001). Competitive response hierarchies for germination, growth, and survival and their influence on abundance. Ecology 82(4): 979-990.

Hubbell, S. P. (2001). The unified neutral theory of biodiversity and biogeography. Princeton, New Jersey, Princeton University Press.

Hubbell, S. P. (2005). Neutral theory in community ecology and the hypothesis of functional equivalence. Funct. Ecol. 19: 166-172.

Huston, M. A. (1999). Local processes and regional patterns: Appropriate scales for understanding variation in the diversity of plants and animals. Oikos 83(3): 393-401.

Hutchinson, G. E. (1951). Copepodilogy for the ornithologist. Ecology 32: 571-577.

Hutchinson, G. E. (1957). Concluding remarks. Cold Spring Harb. Symp. Quant. Biol 22: 415-427.

Hutchinson, G. E. (1959). Homage to Santa Rosalia, or Why are there so many kinds of animals? Am. Nat. 93(870): 145-159.

Jäger, E. J. and K. Werner (2002). Rothmaler Exkursionsflora von Deutschland. Gefäßpflanzen: Kritischer Band. Heidelberg Berlin, Spektrum Akademischer Verlag.

Janssens, F., A. Peeters, J. R. B. Tallowin, J. P. Bakker, R. M. Bekker, F. Fillat and M. J. M. Oomes (1998). Relationship between soil chemical factors and grassland diversity. Plant Soil 202(1): 69-78.

Janzen, D. (1986). The eternal external threat. In: M. E. Soulé. Conservation biology: The science of scarcity and diversity, Sinauer Associates: 286-303.

Jenkins, D. G. (2006). In search of quorum effects in metacommunity structure: Species co-occurrence analyses. Ecology 87(6): 1523-1531.

Jensen, K. and C. Meyer (2001). Effects of light competition and litter on the performance of Viola palustris and on species composition and diversity of an abandoned fen meadow. Plant Ecology 155(2): 169-181.

Jentsch, A. and W. Beyschlag (2003). Vegetation ecology of dry acidic grasslands in the lowland area of central Europe. Flora (Jena) 198: 3-26.

Johst, K., R. Brandl and S. Eber (2002). Metapopulation persistence in dynamic landscapes: The role of dispersal distance. Oikos 98(2): 263-270.

Kalamees, R. and M. Zobel (2002). The role of the seed bank in gap regeneration in a calcareous grassland community. Ecology 83(4): 1017-1025.

Kattwinkel, M., B. Strauss, R. Biedermann and M. Kleyer (2009). Modelling multi-species response to landscape dynamics: Mosaic cycles support urban biodiversity. Landscape Ecol. 24: 929-941.

Keddy, P. A. (1992). Assembly and response rules: Two goals for predictive community ecology. J. Veg. Sci. 3: 157-164.

Keymer, J. E., P. A. Marquet, J. X. Velasco-Hernández and S. A. Levin (2000). Extinction thresholds and metapopulation persistence in dynamic landscapes. Am. Nat. 156(5): 478-494.

Kisdi, É. and S. A. H. Geritz (2003). On the coexistence of perennial plants by the competition-colonization trade-off. Am. Nat. 161(2): 350-354.

Klausmeier, C. A. (2001). Habitat destruction and extinction in competitive and mutualistic metacommunities. Ecol. Lett. 4: 57-63.

Kleyer, M. (1999). Distribution of plant functional types along gradients of disturbance intensity and resource supply in an agricultural landscape. J. Veg. Sci. 10(5): 697-708.

Kleyer, M. (2002). Validation of plant functional types across two contrasting landscapes. J. Veg. Sci. 13: 167-178.

Kleyer, M., R. M. Bekker, I. C. Knevel, J. P. Bakker, K. Thompson, M. Sonnenschein, P. Poschlod, J. M. van Groenendael, L. Klimeš, J. Klimešová, S. Klotz, G. M. Rusch, M. Hermy, D. Adriaens, G. Boedeltje, B. Bossuyt, P. Endels, L. Götzenberger, J. G. Hodgson, A.-K. Jackel, A. Dannemann, I. Kühn, D. Kunzmann, W. A. Ozinga, C.

Römermann, M. Stadler, J. Schlegelmilch, H. J. Steendam, O. Tackenberg, B. Wilmann, J. H. C. Cornelissen, O. Eriksson, E. Garnier, A. Fitter and B. Peco (2008). The LEDA Traitbase: A database of life-history traits of the Northwest European flora. J. Ecol. 96: 1266-1274.

Kleyer, M., R. Biedermann, K. Henle, H.-J. Poethke, P. Poschlod, B. Schröder, J. Settele and D. Vetterlein (2007). Mosaic cycles in agricultural landscapes of Central Europe. Basic Appl. Ecol. 8: 295-309.

Klimeš, L., J. Klimešová, R. Hemdriks and J. M. van Groenendael (1997). Clonal plant architechture: A comparative analysis of form and function. In: H. de Kroon and J. M. van Groenendael. The ecology and evolution of clonal plants. Leiden, Backhuys Publishers.

Klimešová, J. and L. Klimeš (2006). Clo-Pla 3 – database of clonal growth of plants from Central Europe. http://clopla.butbn.cas.cz/.

Kolasa, J. and T. N. Romanuk (2005). Assembly of unequals in the unequal world of a rock pool metacommunity. In: M. Holyoak, M. A. Leibold and R. D. Holt. Metacommunities. Spatial dynmics and ecological communities. Chicago, The University of Chicago: 212-232.

Kolb, A. and M. Diekmann (2005). Effects on life-history traits on responses of plant species to forest fragmentation. Conserv. Biol. 19(3): 929-938.

Körner, C., J. Stöcklin, L. Reuther-Thiébaud and S. Pelaez-Riedl (2008). Small differences in arrival time influence composition and productivity of plant communities. New Phytol. 177(3): 698-705.

Kraft, N. J. B., R. Valencia and D. D. Ackerly (2008). Functional traits and niche-based tree community assembly in an Amazonian forest. Science 322(5901): 580-582.

Krauss, J., A. M. Klein, I. Steffan-Dewenter and T. Tscharntke (2004). Effects of habitat area, isolation, and landscape diversity on plant species richness of calcareous grasslands. Biodivers. Conserv. 13(8): 1427-1439.

Laliberté, E. and P. Legendre (2010). A distance-based framework for measuring functional diversity from multiple traits. Ecology 91: 299-305.

Larcher, W. (1984). Ökologie der Pflanzen. Stuttgart, Verlag E. Ulmer (UTB 232).

Lavorel, S. and E. Garnier (2002). Predicting changes in community composition and ecosystem functioning from plant traits: Revisiting the Holy Grail. Funct. Ecol. 16(5): 545-556.

Lavorel, S., J. Lepart, M. Debussche, J.-D. Lebreton and J.-L. Beffy (1994). Small scale disturbances and the maintenance of species diversity in Mediterranean old fields. Oikos 70(3): 455-473.

Lavorel, S., S. McIntyre and K. Grigulis (1999). Plant response to disturbance in a Mediterranean grassland: How many functional groups? J. Veg. Sc 10(5): 661-672.

Lavorel, S., B. Touzard, J. D. Lebreton and B. Clement (1998). Identifying functional groups for response to disturbance in an abandoned pasture. Acta Oecol. 19(3): 227-240.

Lawrence, R. L. and W. J. Ripple (2000). Fifteen years of revegetation of Mount St. Helens: A landscape-scale analysis. Ecology 81: 2742-2752.

Lawton, J. H. (1999). Are there general laws in ecology? Oikos 84(2): 177-192.

Legendre, P. and L. Legendre (2006). Numerical ecology. Amsterdam, Elsevier.

Leibold, M. A. (1998). Similarity and local co-existence of species in regional biotas. Evol. Ecol. 12(1): 95-110.

Leibold, M. A., M. Holyoak, N. Mouquet, P. Amarasekare, J. M. Chase, M. F. Hoopes, R. D. Holt, J. B. Shurin, R. Law, D. Tilman, M. Loreau and A. Gonzalez (2004). The metacommunity concept: A framework for multi-scale community ecology. Ecol. Lett. 7(7): 601-613.

Leibold, M. A. and M. A. McPeek (2006). Coexistence of the niche and neutral perspectives in community ecology. Ecol. Soc. 87(6): 1399-1410.

Leibold, M. A. and G. M. Mikkelson (2002). Coherence, species turnover, and boundary clumping: Elements of meta-community structure. Oikos 97(2): 237-250.

Leishman, M. R. (1999). How well do plant traits correlate with establishment ability? Evidence from a study of 16 calcareous grassland species. New Phytol. 141(3): 487-496.

Lepš, J., F. de Bello, S. Lavorel and S. Berman (2006). Quantifying and interpreting functional diversity of natural communities: practical considerations matter. Preslia 78: 481-501.

Levin, S. A. (1974). Dispersion and population interactions. Am. Nat. 108(960): 207-228.

Levine, J. M. (2000). Species diversity and biological invasions: Relating local process to community pattern. Science 288: 852-854.

Levine, J. M. and J. HilleRisLambers (2009). The importance of niches for the maintenance of species diversity. Nature 461(7261): 254-257.

Levins, R. (1969). Some demographic and genetic consequences of environmental heterogeneity for biological control. Bulletin of the Entomological Society of America 15: 237-240.

Lewontin, R. C. (1974). The genetic basis of ecolutionary change. New York, Columbia, University Press.

Liancourt, P., K. Tielbörger, S. Bangerter and R. Prasse (2009). Components of 'competitive ability' in the LHS model: Implication on coexistence for twelve co-occurring Mediterranean grasses. Basic Appl. Ecol. 10(8): 707-714.

Loehle, C. (1998). Height Growth Rate Tradeoffs Determine Northern and Southern Range Limits for Trees. J. Biogeogr. 25(4): 735-742.

Long, Z. T., O. L. Petchey and R. D. Holt (2007). The effects of immigration and environmental variability on the persistence of an inferior competitor. Ecol. Lett. 10: 574-585.

Loreau, M., N. Mouquet and A. Gonzalez (2003a). Biodiversity as spatial insurance in heterogeneous landscapes. PNAS 100(22): 12765-12770.

Loreau, M., N. Mouquet and R. D. Holt (2003b). Meta-ecosystems: A theoretical framework for a spatial ecosystem ecology. Ecol. Lett. 6(8): 673-679.

Losos, J. B. (2000). Ecological character displacement and the study of adaptation. PNAS 97: 5693-5695.

Losure, D. A., B. J. Wilsey and K. A. Moloney (2007). Evenness–invasibility relationships differ between two extinction scenarios in tallgrass prairie. Oikos 116(1): 87-98.

Louette, G. and L. De Meester (2007). Predation and priority effects in experimental zooplankton communities. Oikos 116: 419-426.

MacArthur, R. and R. Levins (1967). The limiting similarity, convergence, and divergence of coexisting species. Am. Nat. 101(921): 377-385.

MacArthur, R. H. and E. O. Wilson (1967). The theory of island biogeography. Princeton, Princeton University Press.

Mackey, R. L. and D. J. Currie (2001). The diversity-disturbance relationship: Is it generally strong and peaked? Ecology 82: 3479-3492.

Mahdi, A., R. Law and A. J. Willis (1989). Large niche overlaps among coexisting plant species in a limestone grassland community. J. Ecol. 77(2): 386-400.

Mahmoud, A. and J. P. Grime (1974). A comparison of negative relative growth rate in shaded seedlings. New Phytol. 73: 1215.

Mal, T. K., L.-D. Jon and L. Lovett-Doust (1997). Time-dependent competitive displacement of Typha angustifolia by Lythrum salicaria. Oikos 79(1): 26-33.

Marañón, T. and P. J. Grubb (1993). Physiological basis and ecological significance of the seed size and relative growth rate relationship in Mediterranean annuals. Funct. Ecol. 7(5): 591-599.

Mason, N. W. H., P. Irz, C. Lanoiselée, D. Mouillot and C. Argillier (2008). Evidence that niche specialization explains species-energy relationships in lake fish communities. J. Animal Ecology 77: 285-296.

Mason, N. W. H., C. Lanoiselée, D. Mouillot, P. Irz and C. Argillier (2007). Functional characters combined with null models reveal inconsistency in mechanisms of species turnover in lacustrine fish communities. Oecologia 153(2): 441-452.

McCune, B. and J. B. Grace (2002). Analysis of ecological communities. Gleneden Beach, Oregon, USA, MjM Software Design.

McGill, Enquist, E. Weiher and M. Westoby (2006). Rebuilding community ecology from functional traits. Trends Ecol. Evol. 21(4): 178-184.

Messier, J., B. J. McGill and M. J. Lechowicz (2010). How do traits vary across ecological scales? A case for trait-based ecology. Ecol. Lett. 13(7): 838-848.

Meyer, K. M., K. Wiegand and D. Ward (2009). Patch dynamics integrate mechanisms for savanna tree-grass coexistence. Basic Appl. Ecol. 10: 491-499.

Milberg, P. (1995). Soil seed bank after eighteen years of succession from grassland to forest. Oikos 72(1): 3-13.

Miles, J. (1987). Vegetation succession: Past and present perceptions. In: A. J. Gray, M. J. Crawley and P. J. Edwards. Colonization, succession and stability. Oxford London Edinburgh Boston Palo Alto Melbourne, Blackwell Scientific Publications: 1-30.

Moilanen, A. and M. Nieminen (2002). Simple connectivity measures in spatial ecology. Ecology 83(4): 1131-1145.

Mouquet, N., P. Leadley, J. Meriguet and M. Loreau (2004). Immigration and local competition in herbaceous plant communities: A three-year seed-sowing experiment. Oikos 104: 77-90.

Mouquet, N. and M. Loreau (2003). Community patterns in source-sink metacommunities. Am. Nat. 162(5): 544-557.

Mouquet, N., J. L. Moore and M. Loreau (2002). Plant species richness and community productivity: Why the mechanism that promotes coexistence matters. Ecol. Lett. 5: 56-66.

Muratet, A., N. Machon, F. Jiguet, J. Moret and E. Porcher (2007). The role of urban structures in the distribution of wasteland flora in the Greater Paris Area, France. Ecosystems 10(4): 661-671.

Murphy, J. and J. P. Riley (1962). A modified single solution method for the determination of phosphate in natural waters. Anal. Chim. Acta 27: 31-36.

Murrell, D. J. and R. Law (2003). Heteromyopia and the spatial coexistence of similar competitors. Ecol. Lett. 6: 48-59.

Mwangi, P. N., M. Schmitz, C. Scherber, C. Roscher, J. Schumacher, M. Scherer-Lorenzen, W. W. Weisser and B. Schmid (2007). Niche pre-emption increases with species richness in experimental plant communities. J. Ecol. 95(1): 65-78.

Naeem, S., J. M. H. Knops, D. Tilman, K. M. Howe, T. Kennedy and S. Gale (2000). Plant diversity increases resistance to invasion in the absence of covarying extrinsic factors. Oikos 91(1): 97-108.

Nathan, R. (2006). Long-distance dispersal of plants. Science 313: 786-788.

Nathan, R., F. M. Schurr, O. Spiegel, O. Steinitz, A. Trakhtenbrot and A. Tsoar (2008). Mechanisms of long-distance seed dispersal. Trends Ecol. Evol. 23(11): 638-647.

Navas, M. L. and C. Violle (2009). Plant traits related to competition: How do they shape the functional diversity of communities? Community Ecology 10(1): 131-137.

Nee, S. and G. Stone (2003). The end of the beginning for neutral theory. Trends Ecol. Evol. 18(9): 433-434.

Oksanen, J., R. Kindt, P. Legendre, B. O'Hara, G. L. Simpson, P. Solymos, M. H. H. Stevens and H. Wagner (2009). Vegan: Community ecology package.

Ozinga, W., R. Bekker, J. Schminée and J. M. van Groenendael (2004). Dispersal potential in plant communities depends on environmental conditions. J. Ecol. 92: 767-777.

Ozinga, W. A. (2008). Assembly of plant communities in fragmented landscapes: The role of dispersal. PhD work, Radboud University Nijmegen.

Ozinga, W. A., C. Römermann, R. M. Bekker, A. Prinzing, W. L. M. Tamis, J. H. J. Schaminée, S. M. Hennekens, K. Thompson, P. Poschlod, M. Kleyer, J. P. Bakker and J. M. van Groenendael (2009). Dispersal failure contributes to plant losses in NW Europe. Ecol. Lett. 12(1): 66-74.

Ozinga, W. A., J. H. J. Schaminée, R. M. Bekker, S. Bonn, P. Poschlod, O. Tackenberg, J. Bakker and J. M. van Groenendal (2005). Predictability of plant species composition from environmental conditions is constrained by dispersal limitation. Oikos 108: 555-561.

Pacala, S. W. and M. Rees (1998). Field experiments that test alternative hypotheses explaining successional diversity. Am. Nat. 152: 729-737.

Pacala, S. W. and D. Tilman (1994). Limiting similarity in mechanistic and spatial models of plant competition in heterogeneous environments. Am. Nat. 143(2): 222-257.

Pandit, S. N., J. Kolasa and K. Cottenie (2009). Contrasts between habitat generalists and specialists: An empirical extension to the basic metacommunity framework. Ecology 90(8): 2253-2262.

Patterson, B. D. and W. Atmar (1986). Nested subsets and the structure of insular mammalian faunas and archipelagos. In: L. R. Heaney and B. D. Patterson. Island biogeography of mammals. London, UK, Academic Press: 65-82.

Peet, R. K. (1992). Community structure and ecosystem function. In: D. C. Glenn-Lewin, R. K. Peet and T. T. Veblen. Plant succession: Theory and prediction. London, Chapmann & Hall: 103-151.

Peres-Neto, P. R., J. D. Olden and D. A. Jackson (2001). Environmentally constrained null models: Site suitability as occupancy criterion. Oikos 93: 110-120.

Perry, G. L. W., N. J. Enright, B. P. Miller, B. B. Lamont and R. S. Etienne (2009). Dispersal, edaphic fidelity and speciation in species-rich Western Australian shrublands: Evaluating a neutral model of biodiversity. Oikos 118(9): 1349-1362.

Petchey, O. L. and K. J. Gaston (2002). Functional diversity (FD), species richness and community composition. Ecol. Lett. 5(3): 402-411.

Pianka, E. R. (1966). Latitudinal gradients in species diversity: A review of concepts. Am. Nat. 100: 33-46.

Pianka, E. R. (1980). Guild structure in desert lizards. Oikos 35: 194-201.

Piessens, K., O. Honnay and M. Hermy (2005). The role of fragment area and isolation in the conservation of heathland species. Biol. Conserv. 122(1): 61-69.

Pimm, S. and P. Raven (2000). Biodiversity - Extinction by numbers. Nature 403(6772): 843-845.

Pimm, S. L., G. J. Russell, J. L. Gittleman and T. M. Brooks (1995). The Future of biodiversity. Science 269(5222): 347-350.

Poos, M. S., S. C. Walker and D. A. Jackson (2009). Functional-diversity indices can be driven by methodological choices and species richness. Ecology 90(2): 341-347.

Prach, K. and P. Pyšek (1999). How do species dominating in succession differ from others? J. Veg. Sci. 10(3): 383-392.

Prieur-Richard, A. H., S. Lavorel, K. Grigulis and A. D. Santos (2000). Plant community diversity and invasibility by exotics: Invasion of Mediterranean old fields by "Conyza bonariensis" and "Conyza canadensis". Ecol. Lett. 3(5): 412-422.

Prugh, L. R., K. E. Hodges, A. R. E. Sinclair and J. S. Brashares (2008). Effect of habitat area and isolation on fragmented animal populations. PNAS 105: 20770-20775.

Pulliam, H. R. (1988). Sources, sinks, and pupulation regulation. Am. Nat. 132: 652-661.

Questad, E. J. and B. L. Foster (2008). Coexistence through spatio-temporal heterogeneity and species sorting in grassland plant communities. Ecol. Lett. 11(7): 717-726.

Quétier, F., A. Thébault and S. Lavorel (2007). Plant traits in a state and transition framework as markers of ecosystem response to land-use change. Ecol. Monogr. 77(1): 33-52.

R Development Core Team (2005). A language and environment for statistical computing. Vienna, Austria, R Foundation for Statistical Computing.

Rao, C. R. (1982). Diversity and dissimilarity coefficients: A unified approach. Theor. Pop. Biol. 21: 24-43.

Raunkiaer, C. (1934). The life Forms of plants and statistical plant geography. Oxford, Clarendon Press.

Rebele, F. (1994). Urban ecology and special features of urban ecosystems. Global Ecol. Biogeogr. Lett. 4(6): 173-187.

Reich, P. B., D. S. Ellsworth, M. B. Walters, J. M. Vose, C. Gresham, J. C. Volin and W. D. Bowman (1999). Generality of leaf trait relationships: A test across six biomes. Ecology 80(6): 1955-1969.

Reich, P. B., M. B. Walters and D. S. Ellsworth (1992). Leaf life-span in relation to leaf, plant, and stand characteristics among diverse ecosystems. Ecol. Monogr. 62(3): 365-392.

Reynolds, H. L., G. G. Mittelbach, T. L. Darcy-Hall, G. R. Houseman and K. L. Gross (2007). No effect of varying soil resource heterogeneity on plant species richness in a low fertility grassland. J. Ecology 95(4): 723-733.

Robinson, J. F. and J. E. Dickerson (1987). Does invasion sequence affect community structure? Ecology 68(3): 587-595.

Roscher, C., H. Beßler, Y. Oelmann, C. Engels, W. Wilcke and E.-D. Schulze (2009). Resources, recruitment limitation and invader species identity determine pattern of spontaneous invasion in experimental grasslands. J. Ecol. 97(1): 32-47.

Ruokolainen, L., E. Ranta, V. Kaitala and M. S. Fowler (2009). When can we distinguish between neutral and non-neutral processes in community dynamics under ecological drift? Ecol. Lett. 12(9): 909-919.

Sale, P. F. (1977). Maintenance of high diversity in coral reef fish communities. Am. Nat. 111: 337-359.

Sanderson, R., M. Eyre and S. Rushton (2005). The influence of stream invertebrate composition at neighbouring sites on local assemblage composition. Freshw. Biol. 50: 221-231.

Schadek, U. (2006). Plants in urban brownfields: Modelling the driving factors of site conditions and of plant functional group occurrence in a dynamic environment. Department of Biology and Environmental Sciences, University of Oldenburg.

Schadek, U., B. Strauss, R. Biedermann and M. Kleyer (2009). Plant species richness, vegetation structure and soil resources of urban brownfield sites linked to successional age. Urban Ecosystems Online First: 115-126.

Schaffers, A. P. and K. V. Sykora (2000). Reliability of Ellenberg indicator values for moisture, nitrogen and soil reaction: A comparison with field measurements. J. Veg. Sci. 11(2): 225-244.

Schippers, P., J. M. van Groenendael, L. M. Vleeshouwers and R. Hunt. (2001). Herbaceous plant strategies in disturbed habitats. Oikos 95: 198-210.

Schlichting, E., H. Blume and K. Stahr (1995). Bodenkundliches Praktikum. Berlin, Blackwell.

Schluter, D. (1990). Species-for-species matching. Am. Nat. 136: 560-568.

Schurr, F., G. Midgley, A. Rebelo, G. Reeves, P. Poschlod and S. I. Higgins (2007). Colonization and persistence ability explain the extent to which plant species fill their potential range. Global Ecol. Biogeogr. Lett. 16 (4): 449-459.

Shipley, B. and M. Parent (1991). Germination responses of 64 wetland species in relation to seed size, minimum time to reproduction and seedling relative growth rate. Funct. Ecol. 5(1): 111-118.

Shipley, B. and R. H. Peters (1990). The allometry of seed weight and seedling relative growth rate. Funct. Ecol. 4(4): 523-529.

Shmida, A. and S. Ellner (1984). Coexistence of plant species with similar niches. Plant Ecol. 58(1): 29-55.

Shmida, A. and M. V. Wilson (1985). Biological determinants of species diversity. J. Biogeogr. 12(1): 1-20.

Simberloff, D. (2004). Community ecology: Is it time to move on? Am. Nat. 163(6): 787-799.

Skellam, J. G. (1951). Random dispersal in theoretical populations. Biometrika 38: 196-218.

Smith, C. C. and S. D. Fretwell (1974). The optimal balance between size and number of offspring. Am. Nat. 108: 499-506.

Solé, R. V., D. Alonso and J. Saldaña (2004). Habitat fragmentation and biodiversity collapse in neutral communities. Ecol. Complex. 1: 65-75.

Soons, M. B., R. Nathan and G. G. Katul (2004). Human effects on long-distance wind dispersal and colonization by grassland plants. Ecology 85(11): 3069-3079.

SPSS Inc. (2006). SPSS. Chicago, USA, SPSS Inc.

Steyerberg, E. W., F. E. Harrell, G. J. J. M. Borsboom, M. J. C. Eijkemans, Y. Vergouwe and J. D. F. Habbema (2001). Internal validation of predictive models: Efficiency of some procedures for logistic regression analysis. J. Clin. Epidemiol. 54(8): 774-781.

Strauss, B. and R. Biedermann (2006). Urban brownfields as temporary habitats: Driving forces for the diversity of phytophagous insects. Ecography 29: 928-940.

Stubbs, W. J. and J. B. Wilson (2004). Evidence for limiting similarity in a sand dune community. J. Ecol. 92(4): 557-567.

Suding, K. N., D. E. Goldberg and K. M. Hartman (2003). Relationships among species traits: Separating levels of response and identifying linkages to abundance. Ecology 84(1): 1-16.

Svenning, J. C., D. Harlev, M. Sorensen and H. Balslev (2009). Topographic and spatial controls of palm species distributions in a montane rain forest, Southern Ecuador. Biodivers. Conserv. 18(1): 219-228.

Symstad, A. J. (2000). A test of the effects of functional group richness and composition on grassland invasibility. Ecology 81(1): 99-109.

Tackenberg, O., P. Poschlod and S. Bonn (2003). Assessment of wind dispersal potential in plant species. Ecol. Monogr. 73(2): 191-205.

Tansley, A. G. (1917). On competition between Galium saxatile L. (G. hercynicum Weig.) and Galium sylvestre Poll. (G. asperum Schreb.) on different types of soil. J. Biol. 5(3/4): 173-179.

Ter Heerdt, G. N. J., G. L. Verweij, R. M. Bekker and J. P. Bakker (1996). An improved method for seed-bank analysis: Seedling emergence after removing the soil by sieving. Funct. Ecol. 10(1): 144-151.

Therneau, T. M., B. Atkinson and B. Ripley (2009). rpart: Recursive Partitioning.

Thioulouse, J., S. Chessel, S. Dolédec and J. M. Olivier (1997). ADE4: A multivariate analysis and graphical display software. Statistics and Computing 7: 75-83.

Thompson, K., J. G. Hodgson, J. P. Grime and M. J. W. Burke (2001). Plant traits and temporal scale: evidence from a 5-year invasion experiment using native species. J. Ecology 89: 1054-1060.

Thompson, K., O. L. Petchey, A. P. Askew, N. P. Dunnett, A. P. Beckerman and A. J. Willis (2009). Little evidence for limiting similarity in a long-term study of a roadside plant community. J. Ecol. 98(2): 480-487.

Tichý, L. and J. Holt (2002). JUICE. Program for management, analysis and classification of ecological data. Program manual.

Tilman, D. (1990). Constraints and tradeoffs: toward a predictive theory of competition and succession. Oikos 58: 3-15.

Tilman, D. (1994). Competition and biodiversity in spatially structured habitats: Spatial theory. Ecology 75(11): 2-16.

Tilman, D. (1997). Community invasibility, recruitment limitation, and grassland biodiversity. Ecology 78(1): 81-92.

Tilman, D. (2004). Niche tradeoffs, neutrality, and community structure: A stochastic theory of resource competition, invasion, and community assembly. PNAS 101(30): 10854-10861.

Tilman, D., C. L. Lehman and C. Yin (1997). Habitat destruction, dispersal, and deterministic extinction in competitive communities. Am. Nat. 149(3): 405-435.

Tilman, D., R. M. May, C. L. Lehman and M. A. Nowak (1994). Habitat destruction and the extinction debt. Nature 371: 65-66.

Tranquillini, W. (1979). Physiological ecology of the Alpine timberline: Tree existence at high altitudes with special references to the European Alps. . Berlin, Heidelberg, New York, Springer Verlag.

Tremlová, K. and Z. Münzbergová (2007). Importance of species traits for species distribution in fragmented landscapes. Ecology 88(4): 965-977.

Tuomisto, H., K. Ruokolainen and M. Yli-Halla (2003). Dispersal, environment, and floristic variation of Western Amazonian forests. Science 299(5604): 241-244.

Turnbull, L. A., M. J. Crawley and M. Rees (2000). Are plant populations seed-limited? A review of seed sowing experiments. Oikos 88(2): 225-238.

Turnbull, L. A., S. Rahm, O. Baudois, S. Eichenberger-Glinz, L. Wacker and B. Schmid (2005). Experimental invasion by legumes reveals non-random assembly rules in grassland communities. J. Ecol. 93(6): 1062-1070.

Vandvik, V. and D. E. Goldberg (2006). Sources of diversity in a grassland metacommunity: Quantifiying the contribution of dispersal to species richness. Am. Nat. 168(2): 157-167.

Verheyen, K., M. Vellend, H. van Calster, G. Peterken and M. Hermy (2004). Metapopulation dynamics in changing landscapes: A new spatially realistic model for forest plants. Ecology 85(12): 3302-3312.

Villéger, S., N. Mason and D. Mouillot (2008). New multidimensional functional diversity indices for a multifaceted framework in functional ecology. Ecology 89: 2290-2301.

Violle, C., E. Garnier, J. Lecoeur, C. Roumet, C. Podeur, A. Blanchard and M.-L. Navas (2009). Competition, traits and resource depletion in plant communities. Oecologia 160(4): 747-755.

Violle, C., M.-L. Navas, D. Vile, E. Kazakou, Fortunell, Hummel and E. Garnier (2007). Let the concept be functional. Oikos.

Vitousek, P. M. (1994). Beyond global warming: Ecology and global change. Ecology 75(7): 1861-1876.

Vitousek, P. M. and L. R. Walker (1987). Colonization, succession and resource availability: Ecosystem-level interactions. In: A. J. Gray, M. J. Crawley and P. J. Edwards. Colonization, succession and stability. Oxford London Edinburgh Boston Palo Alto Melbourne, Blackwell Scientific Publications: 207-224.

von Holle, B. and D. Simberloff (2004). Testing Fox´s assembly rule: does plant invasion depend on recipient community structure? Oikos 105: 551-563.

Vos, C. C., J. Verboom, P. F. M. Opdam and C. J. F. ter Braak (2001). Toward ecologically scaled landscape indices. Am. Nat. 157(1): 24-41.

Wake, D. B. and V. T. Vredenburg (2008). Are we in the midst of the sixth mass extinction? A view from the world of amphibians. PNAS 105: 11466-11473.

Warner, R. R. and P. L. Chesson (1985). Coexistence mediated by recruitment fluctuations: A field guide to the storage effect. Am. Nat. 125(6): 769-787.

Warren, J. and C. Topping (2004). A trait specific model of competition in a spatially structured plant community. Ecol. Model. 180(4): 477-485.

Watkins, A. J. and J. B. Wilson (2003). Local texture convergence: a new approach to seeking assembly rules. Oikos 102 (3): 525 - 532.

Watt, A. S. (1947). Pattern and Process in the Plant Community. J. Ecol. 35(1/2): 1-22.

Weiher, E. and P. Keddy (1999). Ecological assembly rules: Perspectives, advances, retreats. Cambridge, Cambridge University Press.

Weiher, E., A. van der Werf, K. Thompson, M. Roderick, E. Garnier and O. Eriksson (1999). Challenging Theophrastus: A common core list of plant traits for functional ecology. J. Veg. Sci. 10(5): 609-620.

Westoby, M., D. S. Falster, A. T. Moles, P. A. Vesk and I. J. Wright (2002). Plant ecological strategies: Some leading dimensions of variation between species. Annu. Rev. Ecol. Syst. 33: 125-159.

Westoby, M., M. Leishman, H. P. Janice Lord and D. J. Schoen (1996). Comparative ecology of seed size and dispersal. Philosophical Transactions: Biological Sciences 351(1345): 1309-1318.

Whittaker, R. H. (1969). Evolution of diversity in plant communities. In: G. M. Woodwell and H. Smith. Diversity and stability in ecological systems: 178-195.

Whittaker, R. H. (1975). Communities and ecosystems. New York, Macmillan.

Wielgolaski, F. and S. Karlsen (2007). Some views on plants in polar and alpine regions. Rev. Environ. Sci. Biotechnol. 6: 33-45.

Wiens, J. A. (1991). Ecomorphological comparisons of the shrub-desert avifaunas of Australia and North America. Oikos 60: 55-63.

Wilson, D. S. (1992). Complex Interactions in Metacommunities, with implications for biodiversity and higher levels of selection. Ecology 73(6): 1984-2000.

Wilson, E. O. (2002). The future of life. New York, Toronto, Alfred A. Knopf.

Wilson, J. B. (1989). A null model of guild proportionality, applied to stratification of a New Zealand temperate rain forest. Oecologia 80: 263-267.

Wilson, J. B. (1999). Assembly rules in plant communities. In: E. Weiher and P. Keddy. Ecological Assembly Rules, Cambridge University press: 130-164.

Wilson, J. B. (2007). Trait-divergence assembly rules have been demonstrated: Limiting similarity lives! A reply to Grime. J. Veg. Sci. 18: 451-452.

Wilson, J. B. and H. Gitay (1995). Community structure and assembly rules in a dune slack: Variance in richness, guild proportionality, biomass constancy and dominance/diversity relations. Vegetatio 116(2): 93-106.

Wilson, J. B. and S. H. Roxburgh (1994). A demonstration of guild-based assembly rules for a plant community, and determination of intrinsic guilds. Oikos 69(2): 267-276.

Wilson, J. B. and R. J. Whittaker (1995). Assembly rules demonstrated in a saltmarsh community. J. Ecol. 83: 801-807.

Woodruff, D. S. (2001). Declines of biomes and biotas and the future of evolution. PNAS 98: 5471-5476.

Wootton, J. T. (2005). Field parameterization and experimental test of the neutral theory of biodiversity. Nature 433(7023): 309-312.

Wright, I. J., P. K. Groom, B. B. Lamont, P. Poot, L. D. Prior, P. B. Reich, E.-D. Schulze, E. J. Veneklaas and M. Westoby (2004a). Leaf trait relationships in Australian plant species. Funct. Plant Biol. 31: 551-558.

Wright, I. J., P. B. Reich, M. Westoby, D. D. Ackerly, Z. Baruch, F. Bongers, J. Cavenders-Bares, T. Chapin, J. H. C. Cornelissen, M. Diemer, J. Flexas, E. Garnier, P. K. Groom, J. Gulias, K. Hikosaka, B. B. Lamont, T. Lee, W. Lee, C. Lusk, J. J. Midgley, M.-L. Navas, Ü. Niinements, J. Oleksyn, N. Osada, H. Poorter, P. Poot, L. Prior, V. I. Pyankow, C. Roumet, S. C. Thomas, M. G. Tjoelker, E. J. Veneklaas and R. Villar (2004b). The worldwide leaf economics spectrum. Nature 428: 821-827.

Zavaleta, E. S. and K. B. Hulvey (2007). Realistic variation in species composition affects grassland production, resource use and invasion resistance. Plant Ecol. 188(1): 39-51.

Die VDM Verlagsservicegesellschaft sucht für wissenschaftliche Verlage abgeschlossene und herausragende

Dissertationen, Habilitationen, Diplomarbeiten, Master Theses, Magisterarbeiten usw.

für die kostenlose Publikation als Fachbuch.

Sie verfügen über eine Arbeit, die hohen inhaltlichen und formalen Ansprüchen genügt, und haben Interesse an einer honorarvergüteten Publikation?

Dann senden Sie bitte erste Informationen über sich und Ihre Arbeit per Email an *info@vdm-vsg.de*.

Sie erhalten kurzfristig unser Feedback!

VDM Verlagsservicegesellschaft mbH
Dudweiler Landstr. 99
D - 66123 Saarbrücken
www.vdm-vsg.de

Telefon +49 681 3720 174
Fax   +49 681 3720 1749

Die VDM Verlagsservicegesellschaft mbH vertritt

Printed by Books on Demand GmbH, Norderstedt / Germany